美食。簡易快
GOURMET COOKING
Made Easy

陳家廚坊

陳紀臨・方曉嵐 著

前言

　　我們的廚房，由使用柴火，到煤炭、煤油（火水）、煤氣、電爐、電焗爐、石油氣LPG、電鍋、微波爐，到現在的電磁爐及各種電子控制的煮食電器，變化的速度越來越快，使用越來越方便、安全、美觀，也更因節能而減省開支，間接減低二氧化碳的排放，這就是新一代的煮食潮流！

　　隨着人們生活水平的普遍提高，健康、環保生活的觀念日漸深入，無火煮食已漸漸被更多人接受。生活緊張，工作繁忙的人們，一方面希望追求美食，但做法要簡易快捷、少油煙、容易清潔，卻又不失營養和美味，新一代的無火煮食電器，就最能幫到您。我們家使用無火煮食已經很多年，我們在香港和台灣出版的十多本食譜書，書中幾百個菜式，都是用家中電磁爐烹調出來的，所以我們深知使用無火煮食的好處。

　　本書介紹的近四十款菜式和飲品，全部是使用新一代的電器廚具，包括IH電磁爐、電焗爐、電子高速煲、磁應電飯煲等。

以簡單明瞭的方法，讓工作忙碌的您，或廚房新手的您，只需要跟隨我們介紹的準備工作程序，然後按下「Start」，便很容易做出包括多款香港人熟悉的日常菜式，以至很多人都絕對意想不到的「溏心鮑魚」、「冬菇扣花膠」等請客菜式！

我們的生活已經進入了電子的世界，潮流追求高效、方便、環保，一個電器化的廚房，一本簡單易學的好食譜，使烹調煮食增添新樂趣，定能為您和家人帶來無限驚喜！

謹以本書，向一直支持我們陳家廚坊系列的讀者們，致以誠懇的謝意！讓我們一起攜手進入烹調的新天地，享受美食，追求健康，身心愉快！

陳紀臨，方曉嵐

2015年秋

目錄

CONTENTS

五味手撕雞

炎炎夏日裏的最佳開胃菜式

　　「五味手撕雞」又名「涼拌手撕雞」，是我們家常做的涼菜，也是父親特級校對最喜歡吃的菜式。父親在1953年出版的十冊《食經》中，詳細記載了他戰時在廣西吃過的「五味手撕雞」。父親對這道涼菜的喜愛，除了味道酸甜可口之外，還有他對那段烽火歲月的難忘回憶，彷彿內心五味紛陳，感慨良多。

　　我們家的做法是買一隻雞回來，一半做五味手撕雞，另一半用來煲湯或煮煲仔飯，一雞兩味，兩個人份量剛剛好！

材料

雞（約600克）1/2 隻　　　鹽1茶匙

白芝麻1茶匙　　　　　　酸薑40克

酸蕎頭5粒　　　　　　　青瓜1/2條

醬汁

生抽1茶匙　　　　　　　糖1茶匙

浙醋1茶匙　　　　　　　辣椒油1/2茶匙

芥末1茶匙　　　　　　　麻油1茶匙

份量
2-4人份

醃製時間
30分鐘

烹調時間
30分鐘

做法

① 把雞用鹽抹勻，醃30分鐘。

② 在鍋內煮沸一鍋水，把雞放下，熄火，浸15分鐘，取出。

③ 待雞放涼後，起出雞肉，用手把雞肉和雞皮撕成長條。

④ 芝麻用白鑊烘乾，備用。

⑤ 酸薑和酸蕎頭切成絲，放入雞絲中拌勻。

⑥ 青瓜洗淨切絲，備用。

⑦ 把生抽、糖、浙醋、辣椒油、芥末、麻油拌勻成醬汁。

⑧ 食前把醬汁淋在雞肉上，拌上青瓜絲，撒上白芝麻即成。

小貼士

① 蕎頭是薤白的頭，酸蕎頭是加米醋和糖浸漬而成。

② 芥末可用黃芥末，也可以用日本芥末。

③ 青瓜絲要最後才拌入，因為容易出水。

蕎頭

肉鬆豆腐蒸水蛋

懷念那些年，媽媽的拿手小菜！

「味道」是一種很抽象的感覺，同一種食品，每個人的感受都不一定相同。比較理性的人說，這是因為每個人的舌頭都不同，浪漫的人說，味道是一種記憶，不知不覺地、深深地印在腦海中，每次吃食物，腦海中的千千萬萬記憶體，就會迅速地遊走一次，尋回那曾經的記憶，那怕是已經印象模糊，大腦都會有反應，告訴自己，這是好吃還是不好吃。

小時候，生活簡單，無憂無慮，上學、嬉戲、吃東西、發白日夢，無論發生甚麼事情都有父母擔當。長大後，一切喜怒哀樂，甚麼事情都要自己搞定，所以總是覺得童年時光是最美好的，當然，還有媽媽親手做的蒸水蛋，那是抹不走的味道記憶！

份量
2-4人份

準備時間
10分鐘

烹調時間
10分鐘

肉鬆豆腐蒸水蛋

材料

嫩豆腐（約250克）1件　　　榨菜10克

雞蛋2隻　　　　　　　　　清雞湯250毫升

乾肉鬆3湯匙　　　　　　　鹽1/2茶匙

葱（切葱花）2條

做法

① 嫩豆腐切約1.5厘米方粒，灑上1/4茶匙鹽拌勻，醃10分鐘，
　 瀝乾水分。

② 榨菜浸水10分鐘後剁碎。

③ 雞蛋打勻，用網篩濾過，加入清雞湯和1/4茶匙鹽拌勻成蛋
　 漿。

④ 把豆腐放在蒸碟中，加入蛋漿，撒上榨菜粒，蓋上微波爐保
　 鮮紙，待水沸後隔水蒸8至10分鐘至蛋熟，取出。

⑤ 趁熱灑上肉鬆和葱花即成。

小貼士

① 在電磁爐或火爐蒸8至10分鐘即可，是待鍋中水沸後放
　 入蒸碟，如果用的是電蒸爐，記得加上足夠的預熱時
　 間。

② 蒸蛋時間的掌握，在於所用蒸碟形狀的深淺，用較深
　 的碟，就要多蒸一兩分鐘，水蛋才會熟。

碗蒸臘味蘿蔔糕

那些年，家裏過年蒸的蘿蔔糕

　　人人愛吃蘿蔔糕，一般都要等到過年期間，其實中秋之後，白蘿蔔已當造，臘味也上了市，隨時可做臘味蘿蔔糕。我們為女兒做的私人訂製小碗蘿蔔糕，一次過做3碗，每碗份量剛剛好吃一次，吃時原碗蒸熱或微波爐叮熱一下就可以了，餘下的就放在雪櫃中隨時可吃，這樣吃得健康又方便。我覺得這個主意真不錯，特意在此介紹給大家。

材料

白蘿蔔600克

廣東臘腸1.5條

生粉1.5湯匙

胡椒粉1/8茶匙

小蝦米20克

粘米粉3湯匙

鹽1/2茶匙

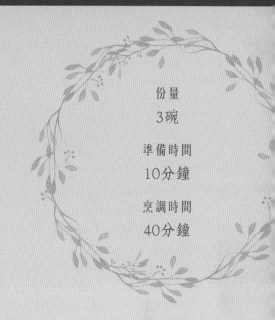

份量
3碗

準備時間
10分鐘

烹調時間
40分鐘

做法

① 把白蘿蔔削皮切成粗絲,放在煲內用慢火把蘿蔔煮至透明,取出放涼。

② 小蝦米洗淨,用水泡軟後剁碎。

③ 臘腸蒸軟或焓軟,切成小粒。

④ 用1湯匙油把蝦米和臘腸炒香。

⑤ 把粘米粉、生粉和蘿蔔拌勻,加鹽、胡椒粉、蝦米和臘腸連油拌勻成蘿蔔糕漿。

⑥ 把蘿蔔糕漿分裝在三個小碗中,用大火蒸30分鐘即成。

小貼士

① 白蘿蔔的含水量會跟隨它的新鮮程度而不同。如果蘿蔔的含水量較少,可酌量加入少量的水。

② 粘米粉與生粉的比例是2:1,如果想蘿蔔糕的口感再硬身一點,可加多一點粘米粉。

菠蘿咕嚕肉

教你美味傳統菜的簡易做法

在世界上任何一個有華人聚居的地方，差不多所有中國餐館都有賣Sweet and Sour Pork，中文叫做咕嚕肉，用的是沒有骨的腬頭肉，適合外國人吃；如果用的是排骨，就叫做甜酸排骨或生炒排骨，其實做法基本上相同。

餐館做這個菜，一般是預先把腬頭肉或排骨用脆粉預先炸好，當有客人下單時，再把炸過的肉翻炸，然後把青椒、菠蘿加進去同炒，最後大火埋個甜酸芡即可上桌。一般家庭很少做這道菜，因為要開炸粉和大油鑊，其實在家中做這道菜，炸過的肉很快就會上桌，是完全沒有必要翻炸的。

這裏介紹的是一個半煎炸的簡易方法，用油量少，豬肉不會被厚厚的粉包着，甜酸味適中。重點是豬肉先用清水加鹽把肉泡過，讓水分滲透到肉裏，肉的口感會更鬆軟。

「澄麵」的廣東話發音是「鄧」麵，是完全除去麵筋的麵粉。因為沒有麵筋，煎炸的時候，熱油較容易進入肉中，能快速地把肉炸得酥透。如果家中沒有澄麵，也可以用生粉或麵粉，但粉層會較厚實。

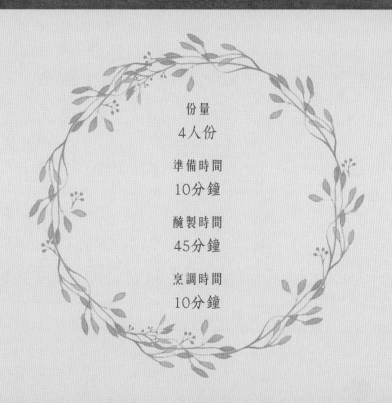

份量
4人份

準備時間
10分鐘

醃製時間
45分鐘

烹調時間
10分鐘

材料

豬脢頭肉300克	鹽1湯匙
洋葱1/2個	青甜椒1/2個
紅甜椒1/2個	罐頭菠蘿（小）1罐
蒜蓉1湯匙	生抽1/2湯匙
酒1/2茶匙	糖1/2茶匙
雞蛋1隻	澄麵4湯匙

糖醋汁材料

大紅浙醋4湯匙	紅糖4湯匙

菠蘿咕嚕肉

做法

① 脢頭肉切成小塊，用500毫升水加1湯匙鹽泡30分鐘，瀝乾水分。

② 洋葱切成塊，青甜椒、紅甜椒和菠蘿切成小塊，罐頭裏的水不要。

③ 用一個碗把浙醋和紅糖拌勻至紅糖完全融化成糖醋汁。

④ 豬肉瀝乾水分後，拌入蒜蓉、生抽、酒和糖，醃15分鐘。

⑤ 把雞蛋打勻，和豬肉拌勻。

⑥ 再用澄麵把豬肉拌勻，使豬肉沾滿粉。

⑦ 用中小火燒熱250毫升油，把豬肉煎炸至金黃，盛起瀝油。

⑧ 倒出鑊內的油，只留1湯匙，把洋葱炒到軟身，放入糖醋汁，炒至汁開始變稠，加入甜椒快炒片刻，放入豬肉，炒至糖醋附在豬肉上，最後放入菠蘿炒勻即成。

小貼士

① 可選用罐頭菠蘿或新鮮菠蘿，或用士多啤梨（草莓）也可以。

② 這道菜要做到油潤而乾身，碟中沒有多餘汁水為高標準。紅糖加熱後有黏性，自然會令味道附在食材上，如果埋芡的話，碟底會留有汁，效果稍遜。

班尼迪蛋

一道充滿挑戰又富有滿足感的簡餐

很多人都以為班尼迪蛋是來自英國的,其實這是一道經典的美國早餐菜式,不過是用了英式鬆餅(English muffin),再加上火腿、雞蛋和荷蘭醬。傳統是一份兩隻蛋,絕對是美式早餐的份量,吃得令人心滿意足。

做班尼迪蛋有兩個主要的步驟,一是做荷蘭醬(Hollandaise sauce),二是做水波蛋(poached egg)。荷蘭醬是班尼迪蛋的靈魂,傳統的做法是用手攪拌雞蛋黃和牛油至完全混合,要求的技術比較高,新手下廚失敗的機會很高。我們提議利用電動手提攪拌機,增加成功的機會。第二是要做好水波蛋,首先是要用一個較大的煲,水要多放,攪拌時漩渦不要太大,盛雞蛋最好用有耳的杯。

英式鬆餅(English muffin)麵包皮比較韌,沾了雞蛋液或火腿汁也不容易散開。火腿的品種很隨意,用普通早餐火腿或巴馬火腿都可以。

材料

雞蛋4隻

英式鬆餅2件

火腿4片

鹽1/2茶匙

白醋1湯匙

荷蘭醬材料

蛋黃2隻

鹽1克

糖1克

檸檬汁1/2湯匙

無鹽牛油100克

份量
2-4人份

烹調時間
20分鐘

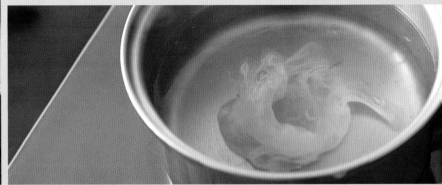

① 把英式鬆餅切開為兩邊，搽上少許牛油，連同火腿片一同放入平底鑊內煎香後取出。

② 把英式鬆餅放在碟中，再在鬆餅上鋪上火腿。

③ 把一隻雞蛋打開，放在杯中，如果蛋黃散開了，便要換另一隻雞蛋。

④ 在鍋內燒開約1.5公升水，加入鹽和醋，轉小火，用長匙在熱水內攪動數次，待水中形成小漩渦後，輕輕將一隻雞蛋順水流方向滑入漩渦中間，讓雞蛋在熱水中浸3-4分鐘至所需生熟程度，將雞蛋撈起，放在廚紙上吸乾水分。

⑤ 把鍋內的水再燒開，用同樣方法處理其他雞蛋。

⑥ 把雞蛋放在火腿上，淋上荷蘭汁即成。

荷蘭醬做法

① 先把牛油煮溶，用網篩把牛油中的牛奶固體過濾，再把濾清的牛油放在量杯裏。

② 把蛋黃、鹽、糖和檸檬汁放到手提攪拌機附件的高身圓筒中，用手提攪拌機把材料打勻，同時慢慢把熱牛油分批從量杯中倒入，邊倒，邊攪拌至牛油和蛋黃漿完全混合。攪拌時可以把攪拌機上下移動，使牛油和蛋黃漿能夠更容易混合。

③ 荷蘭醬做好後，倒出在碗中，蓋住，盡量不要讓醬汁冷卻。

小貼士

① 製作荷蘭醬時要注意，每次加入牛油後要攪拌至完全混和後，才可以加入第二次。

② 白醋能使蛋白迅速凝固，如果白醋的份量不夠，蛋白便會散開。

③ 煮雞蛋時，只需要把水攪拌成小漩渦後，就可以放下雞蛋。如果漩渦太大，容易把雞蛋白沖散。

④ 放下雞蛋時，杯子要盡量靠近水面，所以要用有耳的杯，讓雞蛋順着水流動方向慢慢滑進漩渦中間。攪拌水的時候要注意，如果用左手倒入蛋液，水流要逆時針方向，用右手倒入蛋液，則水流要順時針。

⑤ 做好後，也可再放入焗爐，用面火烤至金黃色，更加美觀和香口，但要小心留意以免烤焦，或導至雞蛋過熟。

韓式泡菜煎餅

半夜肚餓的時候，
隨時可以動手做來祭五臟廟

材料

韓國大白菜辣泡菜50克	紅蘿蔔10克
葱1根	韓國薄餅粉50克
鹽1/8茶匙	糖1/8茶匙
雞蛋1隻	清水50毫升

份量
1至2人份

準備時間
10分鐘

烹調時間
5分鐘

韓風吹遍全球,韓食大放光彩,煎薄餅本是街頭庶民小吃,近年飛上枝頭變鳳凰,成為潮流美食!韓國是泡菜王國,韓國人家中必備有各種泡菜,每一頓飯都要有泡菜伴食,而韓國婦女必定要懂得製作泡菜。韓國位處北方,冬天難以生產蔬菜,所以古代的韓國人已懂得利用發酵技術保存蔬菜,並加入蒜頭、薑、辣椒等辛辣材料,以及水果、小魚乾、鮮蝦等鮮味食材。色彩鮮艷,美味刺激,但又毫不油膩的辣泡菜,據韓國人說,是對身體非常有益的食物。

說到韓國泡菜,其種類繁多,但首先就會想到最經典的大白菜辣泡菜。我家雪櫃中常備一瓶辣泡菜,有時候吃麵也會放上幾塊,醒醒腸胃。隨時可做的食物就是泡菜煎餅,當「家空物淨」又肚子咕咕叫時,家中只要有辣泡菜和雞蛋就有希望了,當然一定要有粉才可煎成餅,最正宗是用韓國薄餅粉,在韓國超市有售,買一包可煎很多次。如果家中沒有韓國薄餅粉,就可以用麵粉或粟米粉代替,如果加入一半的糯米粉,口感就會更柔軟一些。

做法

① 韓國大白菜辣泡菜切成粗絲。

② 紅蘿蔔去皮,切成絲;葱切粗絲。

③ 把薄餅粉、鹽、糖、雞蛋和清水拌勻成薄餅漿。

④ 把泡菜絲、紅蘿蔔絲、葱絲拌入薄餅漿中。

⑤ 在平底不粘鍋裏燒熱1湯匙油,倒下餡料薄餅漿,用慢火煎至微黃色,翻過來把另外一邊也煎成微黃色,盛出。

⑥ 吃時把煎餅切成8塊,即可享用。

滑蛋蝦仁煎米粉

盡享使用電磁爐煮食的優點

無論是煎麵或煎米粉，要煎得金黃而顏色均勻，一般都需要兩到三次調校火力。電磁爐的優點是熱力可以逐級調校，試到熱力合適，毫無難度。而且電磁爐熱力平均，煎米粉或麵時，不用擔心中央部分會因為火力集中而容易烘焦。

　　滑蛋蝦仁要炒得嫩滑，蛋黃和蛋白必須先分開來打勻，因蛋白可打入空氣，炒蛋的口感才會鬆，蛋黃是油性，空氣打不進去；而「熱油、熄火」更是炒蛋嫩滑的要訣。不要怕麻煩，吃過照我們方法做的炒滑蛋，您就會明白了。

材料

廣東米粉150克　　　　　蝦仁300克

雞蛋4-5隻　　　　　　　鹽1茶匙

油適量

小貼士

① 如果蛋液凝結得太慢或者不凝結，即油溫不夠，可以在炒的時候適當提高火力。

② 炒蛋離鑊後，餘溫會繼續把雞蛋煮熟，所以不用炒得太熟。

　滑蛋蝦仁煎米粉

份量
2人份

準備時間
15分鐘

烹調時間
30分鐘

做法

① 在鍋裏煮沸1公升水，放進米粉和1/2茶匙鹽，加蓋熄火，焗10分鐘後，用筷子弄鬆，撈出米粉瀝乾水分，加入2湯匙油把米粉拌勻。

② 蝦仁汆水至熟，瀝乾水分，備用。

③ 用平底不黏鍋，大火燒熱2湯匙油，放入米粉攤平，改用中火耐心煎香一面，翻過來煎香另外一面，剷出放在大碟上。

④ 把雞蛋白和蛋黃分開，把蛋白打至起泡沫，加入1茶匙油和1/2茶匙鹽，再和蛋黃一起打勻，然後放入熟蝦仁。

⑤ 在冷鑊裏下2湯匙油，大火把油燒至中高溫（約170℃），熄火，把蛋液倒進，用鑊鏟沿同一方向把蛋液一層一層的翻轉。

⑥ 到蛋液九成熟的時候，剷出放在米粉上即成。

豉汁豬頸肉蒸腸粉

傳統食物的精彩變奏！

懶洋洋的星期天，來個簡單易做的中式brunch早午餐，豉汁豬頸肉蒸腸粉就最適合了！

豬頸肉又叫做肉青，瘦中有肥，肥中有瘦，肉質爽滑而不肥膩，適宜烹以濃味的醬料，配上豉汁就是最佳的選擇。廣東人叫白腸粉做豬腸粉，因為形狀似豬腸，不只在粥店可以吃到，超市和粉麵店都可以買到。買了白腸粉如果不是即日吃，可拌入少許油，用橄欖油更好，再包好放在雪櫃中，可留一兩天，不用擔心會變硬，只要放上豬頸肉和豉汁來蒸，口感照樣軟滑，好味又方便。

材料

豬頸肉300克

豆豉2湯匙

糖1茶匙

料酒1/2茶匙

小紅辣椒1隻

白腸粉450克

蒜蓉1湯匙

生抽1湯匙

生粉1茶匙

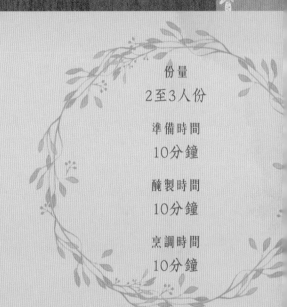

份量
2至3人份

準備時間
10分鐘

醃製時間
10分鐘

烹調時間
10分鐘

做法

① 豆豉搗爛，加入蒜蓉、糖、生抽、料酒及2湯匙水拌勻成豉汁。

② 豬頸肉洗淨，切成肉片，把豉汁放入拌勻，醃10分鐘。

③ 在肉片中加入生粉，蒸前再拌入2湯匙油。

④ 紅辣椒去籽，切成絲。

⑤ 把白腸粉剪成約4至5厘米長段，放入蒸碟中鋪好。

⑥ 把肉片連醃汁放在白腸粉上面，放上紅椒，大火蒸約10分鐘，取出即成。

生煎菜肉鍋貼

上海人的美味秘訣

甚麼叫做生煎？生煎的意思是把全生的食材煎至熟，例如生煎包、生煎鍋貼。生煎的好處是可以保留餡內的肉汁，而同時可以煎得香脆可口。

鍋貼不同煎餅，煎餅是煎兩面，而鍋貼是只煎一面。煎一面但要煎到香脆而豬肉餡料要全熟，就需要有些技巧。

電磁爐的火力，可以隨意逐步減低或升高，而且鍋底的火力比較平均，小火時火力不會集中在中央一小圈，以至有些鍋貼的焦黃程度不同，放在中間的鍋貼餃子較快煎黃，但排在周邊的鍋邊的餃子顏色又未夠。所以，用電磁爐來煎鍋貼，比用一般明火容易控制得多。

如果您懶得自己包鍋貼，可以買超市的各式冷藏餃子。由雪櫃冰格取出，不用解凍，硬冰冰的餃子就可以進行煎，只是煎的時候，加多一點點清水（約多1湯匙），讓煎煮的時間稍為延長，但水不要加得太多，否則會變成水煮，餃子皮會變腍。

材料選購：粉麵店出售的餃子皮，分大小兩種，做鍋貼應選購大餃子皮。

份量
2人份

準備時間
30分鐘

醃製時間
15分鐘

烹調時間
5分鐘

材料

大餃子皮150克	絞豬肉150克
紹菜300克	生抽1茶匙
鹽1/4茶匙	糖1/2茶匙
生粉1/2湯匙	麻油1茶匙
鎮江醋2湯匙	薑絲1湯匙

做法

① 把絞豬肉和生抽、鹽、糖及2湯匙水拌勻，醃15分鐘。

② 紹菜焯至軟身，用冷水沖過瀝水。

③ 把紹菜切碎後再擠出水分，和豬肉拌勻。

④ 加入1/2湯匙生粉，攪拌後再加麻油，拌勻。

⑤ 在餃子皮中間放大約1湯匙餡料，在皮的週邊塗上清水，把皮對摺，先把中間黏住，再把兩邊做成幾個摺黏住，使鍋貼成半月形。

⑥ 用平底易潔鑊，中火燒熱1湯匙油，把鍋貼排好在鑊中，加入60毫升（約4湯匙）清水，煮至水沸，蓋上鑊蓋，改用小火，慢慢煎至水分完全蒸發，再把鍋貼底煎至焦黃即成。

⑦ 上桌時伴以鎮江醋和薑絲。

臘腸滑雞煲仔飯

冬天寒風冽冽，回家弄個暖粒粒的煲仔飯

在家做煲仔飯並不難，但要米飯和配料都熟透，還要烤出一些飯焦（鍋巴）就較難掌握了，竅門是在於控制火候。煮煲仔飯用明火或電磁爐都可以，如果家中廚房是用明火煮食，就要用手動調校火力的大小，但明火的小火，一定是集中在中央一小圈，鍋底的熱力不平均，解決這問題的方法很簡單，到廚房用具店買一塊烘焗煲仔飯用的鐵片，加在明火上，在焗飯時把鍋放上去，這樣火力就平均了；如果家中廚房是用電磁爐煮食，做法就簡單多了，大火開蓋把米飯煮至快收水，加蓋，把火力調至最低火，由於熱力平均，焗飯的時候，設置好時間掣就不用再看管，新手下廚也會是零失敗。

材料

廣東臘腸1條	鮮雞（或雞腿2隻）半隻
白米160毫升	蒜蓉2湯匙
薑汁1湯匙	生抽1茶匙
生粉1湯匙	薑絲1湯匙
葱白2條	

煲仔飯豉油

老抽1湯匙	生抽2湯匙
雞湯或開水1湯匙	糖1/2茶匙
麻油1/2茶匙	

把所有材料混合煮沸或微波爐加熱即成。

份量
2人份

準備時間
10分鐘

醃製時間
20分鐘

烹調時間
30分鐘

做法

① 臘腸洗淨，斜切成1厘米厚片，葱白切成5厘米長段。

② 雞斬件，用生抽、薑汁醃20分鐘，再加生粉拌勻。燒熱1湯匙油，把雞件爆香炒至半熟，盛出備用。

③ 白米洗淨瀝乾，先用1湯匙油爆香蒜蓉，再把米放入一同炒約半分鐘。

④ 把白米放入煲內，加適量水（平日煮飯的同等份量清水），打開蓋用大火煮沸。

⑤ 看到米快要收水，放入臘腸和雞件，鋪上葱白和薑絲，蓋上鍋蓋，轉用小火，焗煮約20至25分鐘即可。

⑥ 吃時淋上煲仔飯醬油。

椰汁咖喱大蝦

顏色濃郁，味道香濃，
令你看著無限的食欲！

材料

大海蝦300克	生粉2湯匙	份量
秋葵6條	蒜頭4瓣	2人份
乾葱4粒	紅尖椒1隻	準備時間
乾辣椒3隻	紅咖喱粉/咖喱醬1湯匙	10分鐘
椰漿50毫升	糖1湯匙	烹調時間
魚露1湯匙	羅勒1株	10分鐘

做法

① 剪去蝦鬚蝦腳，挑出蝦腸，洗淨，從腹部剖開，內外拍上生粉。

② 燒熱4湯匙油，把蝦分批下鍋煎炸至金黃，取出瀝油。

③ 秋葵洗淨，切去蒂，備用。

④ 蒜頭剁蓉，乾蔥切片，紅尖椒切去蒂，開邊去籽。

⑤ 用鑊中餘油，中火爆香蒜頭、乾蔥、乾辣椒，再放入咖喱粉炒香，加入2湯匙水。

⑤ 放入秋葵和紅尖椒同炒。

⑥ 放進大蝦和椰漿、糖、魚露，一起煮至收汁。

⑦ 裝盤後放上羅勒即成。

小貼士

① 乾辣椒作用是增加少許香辣味，不吃辣的人可不加乾辣椒。

② 紅尖椒是大隻的尖辣椒，不是指泰國小紅椒。紅尖椒一般不太辣，作用是增加色彩。

蜜汁叉燒

自己動手，做一碟充滿誘惑的叉燒

叉燒的受歡迎程度極高，歷久不衰！其實製作叉燒，好吃與否，最關鍵的要訣，不在於用普通豬肉或者是黑毛豬，而是要精選豬肉的部位。做蜜汁叉燒，最好選用豬胛頭肉，也稱為枚肉或胛頭，位置在豬的前肩，即豬的頸背肌肉，肉質瘦中帶肥，口感鬆嫩，肉味香濃，甘腴可口，絕對可以做出「肥叉」；近年香港某些名店中菜館，流行用全瘦的胛肉或柳胛肉，做出來的叉燒肉質非常嫩，但欠缺的是傳統叉燒的甘腴和口感；另外有一些燒臘店或餐廳，用冷藏豬腿肉來做叉燒，肉質乾而粗糙，肉味不佳，這種叉燒通常會在飯盒中出現，令人沮喪。

在家自製蜜汁叉燒，可以選用適合自己的不同肥瘦肉質要求，做到「肥叉留香」絕無困難。同時可以避免吃街外叉燒同時帶來的味精雞粉，以及不必要的食用色素。即燒即食，更是靚叉燒重要的美味因素！

份量
4人份

準備時間
15分鐘

醃製時間
90分鐘

烤焗時間
27分鐘

醃料

糖6湯匙	鹽1湯匙
五香粉1茶匙	海鮮醬2湯匙
沙薑粉1/2茶匙	生抽1/2湯匙
紹興酒2湯匙	蒜蓉2湯匙
乾葱蓉2湯匙	薑汁1湯匙

蜜汁料

麥芽糖3湯匙

糖3湯匙

味醂1湯匙

熱開水1湯匙

蜜汁做法

把麥芽糖、糖、味醂和熱開水放小碗內拌勻，用微波爐叮30秒，或放在不銹鋼碗中，碗下用沸水浸至碗熱，不斷拌勻即成。

叉燒做法

① 把豬肉洗淨，切成2至3條約2至2.5cm厚的長形條。

② 在500毫升清水中加入1湯匙鹽拌勻，放入豬肉，水要浸過肉面，浸泡1/2小時，取出，用清水沖洗一下，用紙吸乾水分。

③ 把醃料的所有材料拌勻，放入豬肉醃製1小時，中途翻動兩三次。

④ 把醃好的豬肉用叉燒針順肉的長度穿好。

⑤ 焗爐預熱190℃，在烤盤上放一張鋁箔紙，放上鋼網，再放上豬肉，焗16-17分鐘。

⑥ 打開焗爐，在叉燒表面掃上一層蜜汁，放回電焗爐，改用燒烤火（上火）烤5分鐘，翻動1次，掃上一層蜜汁，再烤5分鐘。

⑦ 取出叉燒，掃上一層蜜汁即成。

小貼士

① 豬肉要選帶肥的脢頭肉，切成長條的肉條，每條也要有一點肥肉，燒出來的叉燒，口感才會鬆軟甘腍。

② 無論你選用哪一種豬肉來自製蜜汁叉燒，調味和烤焗的時間都基本相同。

③ 粵式燒臘會加入玫瑰露酒，我們改為在蜜汁中加入日本味醂，即低度的甜酒，可增加叉燒的亮澤。

④ 掃蜜汁時，要掃厚一些，讓肉的熱力把濃稠的蜜汁慢慢融化，把肉完全包住。

⑤ 如果喜歡叉燒有甜豉油汁，可另煮汁淋上。方法是把剩下的醃汁過濾，加入少許糖和清水煮溶，淋在切好的叉燒上即可。

蜜燒雞膶

您也可以做出甜蜜甘腴的傳統真味

材料

雞膶8副

蜜汁材料

麥芽糖2湯匙

糖2湯匙

味醂1湯匙

其他材料

竹籤4根

醃料

糖6湯匙

五香粉1茶匙

海鮮醬2湯匙

沙薑粉1/2茶匙

生抽1/2湯匙

紹興酒2湯匙

蒜蓉2湯匙

乾葱蓉2湯匙

薑汁1湯匙

做完蜜汁叉燒，剩下一碗醃汁醬料和蜜汁，正好跟著就做個燒雞膶（雞肝）！

雖然不少人都聞膶色變，但喜歡吃靚燒雞膶的人更多，可惜燒臘店或酒樓的燒雞膶，一般都會烤得過了火。傳統叉燒是利用燒鴨爐的爐尾火，行內人說「初爐燒鵝，尾爐叉燒」，那麼，燒雞膶更是「大砲打蒼蠅」。由於烤爐大，一般都會烤得過火，雞膶會因過熟而變乾，所以街上很少吃到好吃的燒雞膶。

在家自己用電焗爐來做燒雞膶就靈活多了！先焗後烤，只要準確控制時間，一定可以烤到熟而不乾。雞膶出爐就要趁熱吃，保證口感嫩滑，甘腴無比，配上一杯紅葡萄酒，保證是人間美食！

份量
4人份

準備時間
10分鐘

醃製時間
30分鐘

烹調時間
20分鐘

做法

① 把雞膶上的脂肪去掉，挑出血管，洗淨瀝乾。

② 雞膶放在大碗內，加入醃料拌勻，醃半小時。

③ 把蜜汁材料加1湯匙熱水拌勻，備用。

④ 把串雞膶用的竹籤用清水泡10分鐘。

⑤ 用竹籤把雞膶串好，每一串用兩根竹籤，每一根竹籤穿過雞膶的一邊，四塊雞膶做1串。

⑥ 焗爐預熱到190℃，在烤盤上放一張鋁箔紙，抹上一層油，把穿好的雞膶放在鋁箔紙上。

⑦ 把烤盤放入焗爐內的中格，焗10分鐘。

⑧ 取出烤盤，在雞膶掃上一層蜜汁，再放2分鐘，讓雞膶稍為涼卻。

⑨ 把烤盤放回焗爐的上格，用燒烤火（上火）烤3分鐘，取出。

⑩ 把雞膶再掃上一層蜜汁，翻過另一面，放回焗爐再用上火烤2分鐘即成。

鹽烤馬友魚

如果你不想煎魚弄到滿屋子油煙味，
試試做鹽烤魚吧！

愛吃日本料理的人，都會吃過日式鹽烤魚，通常會用香魚、雞魚及鯛魚（鯢魚）烤製。其實鹽烤魚，主要取其原汁原味，簡簡單單地烤熟，味道自然鮮美，而且不需要加油，健康有益。

香港的海魚種類很多，做鹽烤魚可以選擇用馬友魚、紅鯢、金絲鯢、鯧魚、沙尖魚、帶魚、海鱸等海魚，但是石斑魚不宜做鹽烤魚，肉質會變韌；馬頭魚、青筋魚、黃花魚等，因為魚肉容易散開，也不大適合直接入烤爐，但可以用鋁箔紙包着來焗。

鹽烤魚中，我們最喜歡吃鹽烤馬友魚。馬友魚味道鮮美，骨刺少，魚肉細潤如酥，是海魚的上乘之選，味道尤勝一般石斑。雖然馬友魚在冰鮮魚中屬於上價魚，但絕對是物有所值。以前香港的馬友魚只在三至五月當造時上市，如今似乎一年四季都有供應，街市魚檔及大型超市都有售。馬友魚離水即死，所以市場上只有冰鮮馬友魚出售。

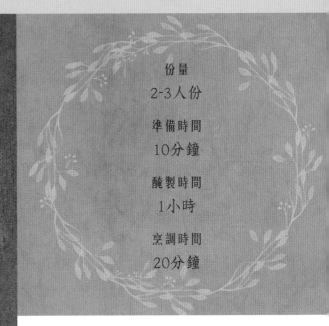

份量
2-3人份

準備時間
10分鐘

醃製時間
1小時

烹調時間
20分鐘

材料

馬友魚1條（約500克）

鹽（醃魚用）1/2茶匙

鹽（尾鰭用）1/2茶匙

胡椒粉1/4茶匙

蛋白1湯匙

做法

① 馬友魚刮鱗宰好，沖洗乾淨。

② 翻開魚腹，用刀把藏在脊骨旁邊的血管劃開，用手指或刷子把血管內的血清除，再用清水沖洗乾淨，用廚紙或毛巾把魚身內外的水分吸乾。

③ 在馬友魚兩面身上淺剞兩三刀，用1/2茶匙鹽把魚身和魚腹抹勻，再撒上胡椒粉，醃1小時。

④ 在馬友魚的尾鰭上沾雞蛋白，再抹一層厚鹽，避免尾鰭烤焦。

⑤ 取出焗盤，鋪上一張鋁箔紙，放上烤魚網（鋼網），然後在網上抹上一層油，放上馬友魚。

⑥ 把焗爐預熱到200℃，把魚放入焗爐，焗8分鐘。

⑦ 把魚翻到另一面，再焗8分鐘。

⑧ 轉用上火烤4分鐘，取出，即可上桌。

蒜香金沙骨

享受啃骨頭加吮手指的私密時光

材料

豬肋排300克	生抽1茶匙
蒜蓉3湯匙	乾葱蓉4湯匙
鹽1/4茶匙	糖1湯匙
生粉1茶匙	咖喱粉1/2湯匙
番茄醬1/2湯匙	牛油2湯匙

　　這是一道典型的「豉油西餐」菜式，五十年代叫做「瑞士排骨」，當時有一間著名的餐廳，推出了這一道菜，據說因為英文的「甜」字與瑞士Swiss的讀音相近，所以乾脆就叫做「瑞士排骨」，事實上此菜與瑞士這個國家並無任何關係。此菜所用的調料，有生抽，也有牛油，有番茄醬，但又有咖喱醬，絕對是中西合璧的國際版，真佩服香港廚師的偉大發明。

　　「蒜香金沙骨」即「瑞士排骨」，餐館的做法是把醃了味的排骨先用油炸或煎好，有客點菜即再放入電焗爐中烤至金黃或快速再炸一次，所以含油量高。為了身體健康着想，我們都希望盡量減少吃煎炸的食物，而減少油煙，更是符合現代廚房的要求。所以，以下介紹的「蒜香金沙骨」，做法經過改良，用電焗爐一次過做成，油分少了，油煙少了，做法簡單了，排骨蒜味香濃，肉質外脆內嫩，好味指數超Like！

份量
2人份

準備時間
20分鐘

烹調時間
20分鐘

做法

① 把排骨斬成6至7塊，每塊約5厘米長，拌放入生抽、蒜蓉、乾葱蓉、鹽、糖和2湯匙水拌勻，醃15分鐘。

② 加入生粉、咖喱粉和番茄醬拌勻。

③ 把排骨排好在烤盤上，多餘的醬汁均勻的放在排骨上，再掃上煮溶了的牛油。

④ 把電焗爐預熱至200℃，放入排骨，焗約10分鐘。

⑤ 把排骨翻過來再焗10分鐘，至排骨表面略焦即成。

芝士焗椰菜花

香噴噴的焗芝士，魅力沒法擋！

當您的廚房設備簡單到不能再簡單，只要您擁有一個小焗爐，您也可以做出千變萬化的中西美食！

電焗爐（烤爐、烤箱）可以做麵包、焗海鮮、烤雞、焗素菜，又可以用來番熱食物，懶惰的您，還可以隔一段時間才清洗它，不用每餐洗鍋洗鑊。

加芝士焗的食物，香噴噴的，令人感到非常滿足。不想食物弄得太複雜，這道「芝士焗椰菜花」就最適合您，週末焗一個來慰勞自己，加一杯白葡萄酒，做個窩心的梳化薯仔吧！

材料

椰菜花（約250）克 1/2個

早餐火腿2片

雞蛋 3隻

芝士白汁

片裝芝士2片

牛油2湯匙

麵粉2湯匙

牛奶250毫升

鹽1/2茶匙

烤面料

帕馬臣芝士碎1湯匙

麵包糠1湯匙

份量
2人份

準備時間
20分鐘

烹調時間
30分鐘

小貼士

① 早餐火腿最容易買到，味道也不太鹹。可選用其他各種火腿，但要注意火腿的鹹度，如果火腿味道重，芝士白汁中的鹽可以減少或取消。

② 如果覺得白汁太稠，可多加50毫升牛奶或水。

做法

① 椰菜花切去硬莖，切成小朵，用水灼熟，瀝乾水分，放在焗盤中。

② 早餐火腿切成碎粒。雞蛋焓熟，剝殼，切成蛋碎。

③ 片裝芝士切碎。

④ 用小火燒溶牛油，加入麵粉，不停攪動煮至完全混合，沒有麵粉團，倒入牛奶，加入鹽，煮至微沸，加入芝士碎，小火煮至材料全部溶化成芝士白汁。

⑤ 把1/3的芝士白汁淋在椰菜花上，拌勻。

⑥ 把火腿和蛋碎平均地灑在椰菜花上，再把餘下的芝士白汁倒在上面。

⑦ 灑上帕馬臣芝士碎和麵包糠。

⑧ 電焗爐預熱至190℃。

⑨ 把焗盤放入電焗爐中，焗15分鐘，然後用上火烤約4分鐘至顏色焦黃即可。

香草焗雞

小焗爐，大用處！

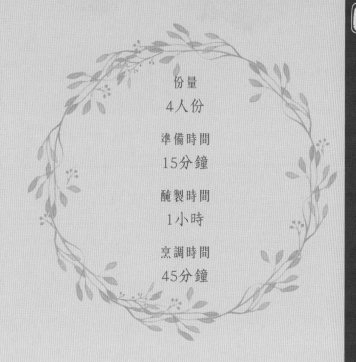

份量
4人份

準備時間
15分鐘

醃製時間
1小時

烹調時間
45分鐘

材料

光雞（淨重約1.2千克）1隻

鹽2茶匙

蒜蓉1.5湯匙

混合香草1.5湯匙

（意大利或普羅旺斯混合香草）

橄欖油2湯匙

馬鈴薯300克

紅蘿蔔150克

小洋蔥200克

不用羨慕豪宅的大廚房，只要您擁有一個電焗爐，就可以輕輕鬆鬆地烘焙蛋糕、焗麵包、烤肉、烤魚、焗芝士意粉……花樣百出，為您平淡的生活增添上一道彩虹。

① 雞洗淨，切去雞頭和頸不要。在雞腿關節下1厘米把雞爪切掉。

② 用鹽把雞的內外抹勻，醃1小時。

③ 把馬鈴薯切開一半，紅蘿蔔切件，洋蔥切大塊。

④ 把蒜蓉、香草和1湯匙橄欖油，混合成香料醬，用1/2份量把雞腔內抹勻。

⑤ 用手指伸進雞胸皮與肉之間，分開皮肉，再把1/4的香料醬塗在雞胸肉上，並用手指輕輕按摩雞肉。

⑥ 把餘下的1/4香料醬抹在雞皮上。

⑦ 用線把雞尾巴的開口縫起來，再用棉繩把兩條雞腿縛在一起，使雞腿向雞身靠攏。

⑧ 兩隻翅膀向後翻，貼在雞背上。

⑨ 預熱電焗爐到190℃。

⑩ 把雞放在烤盤上，旁邊放馬鈴薯、紅蘿蔔、小洋蔥等配菜，再淋上1湯匙橄欖油，放入焗爐焗45分鐘。

⑪ 把雞取出，剪斷棉繩，把縫合雞尾巴的線取出。

⑫ 把雞腔內的汁倒出，放在小碗內作為蘸汁，即可享用。

鱈魚西京燒

魚肉添上味噌香

我們常見到日式料理中有「蒲燒」、「照燒」和「西京燒」等詞語，「蒲燒」和「西京燒」多指燒鰻魚和燒魚，而「照燒」通常就是牛肉。究竟這三者指的是甚麼意思？它們的做法有分別嗎？

蒲燒，是料理中特指烤鰻魚的術語，日本人把鰻魚剖開再攤平，像一塊蒲葉那樣來燒，所以叫做蒲燒，有張開、打開的意思。蒲燒傳到香港，年輕人就理解得更開放，加上中文中的「酺」字，意思是形容飲酒作樂或聚會，「酺」字並不流行，於是就產生了「蒲」，例如「蒲點」和「去邊度蒲？」這些俗語，而「蒲」的音與水泡（粵音水抱）相似，浮來浮去，蒲來蒲去，意思就更形象化了。

照燒，也是日本料理的烹調術語，很多香港人都以為是「照住來燒」，用燈或噴槍來燒的意思。其實據說這個「照」字在日文中是光亮、光澤的意思，在肉類或魚類材料中加入味醂來醃過，再進行燒烤，食物表面就會有光澤，這種做法叫做「照燒」。我們在做「蜜汁叉燒」（見本書第40頁）時也加入味醂，以增加叉燒的光澤。

西京燒，西京是指日本的大坂和京都一帶，江戶時期前，這裏是日本的權力中心。日本的製醬技術源自中國，元朝時期由日本僧人覺心帶到日本，而味噌miso是最具特色醬料，類似我們的麵豉醬、柱侯醬。西京地區生產一種別具風味的白味噌shiro miso用來醃魚然後燒烤，就成了魚肉帶有豉香的西京燒。

白味噌

鱈魚西京燒

份量
2人份

準備時間
10分鐘

醃製時間
1到2天

烹調時間
15分鐘

材料

鱈魚250克

味醂2湯匙

清酒2湯匙

白味噌1.5湯匙

糖1.5湯匙

做法

① 洗淨鱈魚，瀝乾，用廚紙把多餘的水分吸乾。

② 把味醂和清酒放在小鍋內，用慢火煮沸，讓酒精完全揮發。

③ 放入味噌，用勺把味噌、味醂和清酒攪拌至完全混合。

④ 加糖，煮至完全融化。放在碗中，待涼。

⑤ 把魚放在保鮮食物袋內，加入已經涼卻的醬汁，再把食物袋的空氣擠出，封口。

⑥ 用手把醬汁和魚拌勻，輕輕按摩幾下，再放在雪櫃內醃1至2天。

⑦ 把魚從食物袋裏取出，用手把表面上的醬汁抹掉，放至室溫。

⑧ 把焗爐預熱至230℃，放入鱈魚焗約6分鐘，翻過來再焗6分鐘即成。另一選擇是用上火，每一面烤約4分鐘至稍為焦黃即成。

小貼士

鱈魚西京燒，也可以用比目魚或海鱸等海魚來做。

發花膠

花膠是滋補的極品，有養顏護膚的功效，是愛美女士們的恩物。鮑參翅肚中的肚，就是花膠，即大魚的魚鰾，亦即魚體內的空氣囊，作用是調控魚在水中的浮沉。乾貨花膠的顏色金黃而呈半透明，硬實而沒有味道，很像一塊塑膠，所以叫做花膠。

花膠分海魚及淡水魚膠兩大類，價格及其身份，就以其罕有性，以及體形、厚薄來區分。薄花膠產量比較多，價格雖然也不平宜，但要比厚花膠低很多。由於海洋資源被濫捕，上品的花膠供應越來越少，花膠的價格也越來越高。

用電子高速煲煮食，除了能縮短烹調時間之外，更能降低食物中營養素的破壞。用電子高速煲浸發花膠，更加安全快捷，按照我們以下介紹的方法，把浸好的花膠放入電子高速煲中，依方法設定，你就可以去上班，去購物，或者安坐家中看電視和休息，不用擔心任何事。到自動程式完成之後，可即時打開，或回家後再打開，這時花膠已經發好。

發好的花膠，用清水沖洗乾淨，抹乾水分，用三文治食物袋每袋放1隻花膠，擠出空氣，密封好，放入雪櫃的凍格中冷藏。建議一次過發好幾隻花膠，隨時可以拿一兩隻出來做小菜或燉湯，方便快捷。由於發好的花膠，膠質很重，所以每小袋最好只放1隻發好的花膠，以免黏在一起。

要注意的是，如果不是要做厚身的花膠扒，就不需要買名貴的厚身大花膠。建議到海味店選購12至14頭的紮膠公花膠，即每斤（600克）有12至14隻，每隻平均重約50克，這種尺寸和厚度的花膠，很適合用電子高速煲來發製，而且這種花膠價格並不太昂貴，適合喜歡經常吃花膠來補身美容的女士。

材料

12至14頭紮膠公花膠6至7隻
（乾貨重量，平均每隻約50克）

薑（約50克）4片

清水適量

做法

① 把花膠用清水浸泡8小時或1晚，浸至
軟身，用水洗淨。

② 在電子高速煲中放進花膠和薑片，倒
入清水，水要浸過花膠面約4厘米，蓋
上煲蓋。

③ 設置「Low」低壓12分鐘，再按
「Start」開始，待放壓後即完成。如
果是比較厚的花膠，可加多幾分鐘，
至花膠夠軟身。

④ 取出發好的花膠，放入清水中浸洗即
可煮食，或獨立包裝冷凍貯存。

冬菇燴花膠

用電子高速煲自製的一道名貴菜式

花膠發好了，就可以做出很多不同的菜式，發好的花膠可以燉湯、涼拌、切絲炒小菜，但由於花膠的黏性很強，以「燴」製方法來烹調，是最適合不過的。

冬菇燴花膠，可以用以下介紹的電子高速煲方法來做，烹調時間比較短，亦少用了油。準備時間10分鐘，但並不包括用水浸發冬菇，因各種冬菇的品種、厚薄不同，所需的浸發時間都不一樣。

當然，也可以用傳統的方法，先把醃好的冬菇蒸45分鐘至入味，再用中火燒熱2湯匙油，爆香薑片和冬菇，贊酒，加入雞湯、糖、鹽和用傳統方法發好的花膠，燴製8分鐘，加麻油炒勻即成。

材料

花膠（未發重量）50克	冬菇8朵
蠔油1湯匙	生粉1/2湯匙
薑汁1湯匙	清雞湯2湯匙
紹酒1/2湯匙	糖1/2茶匙
鹽1/2茶匙	麻油1/2茶匙

份量
2-4人份

準備時間
10分鐘

烹調時間
15分鐘

做法

① 把花膠發好，用電子高速煲烹調至軟身（見本書第62頁），取出，切成四塊。

② 用水泡軟冬菇，去蒂，擠乾水，拌入蠔油和生粉，再加入1茶匙油拌勻。

③ 把冬菇放入電子高速煲，加清雞湯和薑汁，加蓋，設置「Low」低壓10分鐘，按「Start」。

④ 放壓完成後，打開蓋，加入花膠、酒、糖、鹽一起拌勻。

⑤ 加蓋，設置「Low」低壓5分鐘，按「Start」，完成後拌入麻油，取出裝盤即成。

溏心鮑魚

成為驕人的鮑魚大廚師，並非遙不可及！

　　鮑參翅肚，鮑魚名居榜首，而榜首之首，就是乾鮑。有道是富豪吃大乾鮑，平民百姓吃罐頭鮑，海鮮酒家做鮮鮑。做得好的乾鮑，味道和口感，遠非新鮮鮑魚、急凍鮑魚、罐頭鮑魚所能及。五、六十年代，富貴人家請客，請大酒家廚師來家裏做到會，以示架勢識食，主人家珍藏的兩頭三頭的日本大網鮑就是當然主角，這種頂級大網鮑今時今日極為罕見，當然價值不菲。

　　香港市場上的乾鮑，主要來自日本、南非和中東，其中以日本吉品鮑最受歡迎。日本吉品鮑魚產自日本岩手縣，常見的頭數由12頭至30頭（1斤即600克的隻數）。日本吉品鮑外型呈元寶形，肉身較厚而堅挺，味道鮮美，是做溏心鮑魚的首選，也是中產人家可以接受的價錢。

　　溏心鮑魚是高級中菜館的名貴招牌菜，溏心二字本來是用來形容溏心皮蛋的蛋黃軟糯香滑，溏心鮑魚就是借用溏心來形容鮑魚的軟糯口感。一般家庭很少在家做溏心鮑魚，因為乾鮑的價格昂貴，只怕一旦搞呃了，名貴的鮑魚變成鐵蛋；而且做傳統溏心鮑魚，花的時間很多，光是慢火燜煮就要八至十小時，其間還要小心看管免黏鍋底，實在太費神了！不過，只要您有一個電子高速煲，跟着我們介紹的做法一步步去做，您也可以做出美味的溏心鮑魚！

做溏心鮑魚，必須要注意以下問題：

1. 乾鮑魚怕鹽，首先要把乾鮑的鹽分完全洗去，可以用汆水、浸泡等方法。更重要的是，在煮的過程中，絕不能加鹹味，包括鹽和醬油，否則鮑魚會收縮變硬。所以燜鮑魚要用排骨、江瑤柱等鮮味來入味，就是不能有鹽，包括切忌下火腿。

2. 鮑魚燜好後，要把鮑魚挾出，放在一個架上，盡量讓它四面都接觸到空氣，放置2至3小時，這就是使鮑魚中心部分溏化的過程。

30頭日本吉品鮑魚10隻　　　　　　　　江瑤柱6粒

豬肉腩排400克　　　　　　　　　　　　雞腳12隻

新鮮豬皮（每片約7.5 x 15 cm）2片

薑2片　　　　　　　　　　　　　　　　蔗片糖50克

做法

① 將鮑魚用水洗淨，用清水浸泡一晚。第二天早上將鮑魚水倒去，用軟刷將鮑魚裙邊及鮑魚嘴輕輕洗刷乾淨。

② 把鮑魚放在食物盒中，注入清水，水分要浸過鮑魚面起碼1厘米，封蓋，放在雪櫃2天。

③ 江瑤柱用清水浸泡一個晚上，水留用。

④ 腩排洗淨斬成幾件，雞腳洗淨，豬皮刮去豬毛和脂肪，洗淨，一起放在大鍋中氽水3分鐘，撈起，用清水沖洗。

⑤ 將腩排放在電子高速煲內的最低層，鮑魚、雞腳、江瑤柱、薑片、蔗片糖等順序放入，豬皮鋪在最上面。加入浸鮑魚及江瑤柱的水，加水至覆蓋豬皮面，將煲蓋蓋好。

⑥ 將高速煲設置「High」高壓50分鐘，按「Start」啟動至完成程式。洩壓後，再重複做一次高壓50分鐘，即一共煮兩次。

⑦ 完成後，不要開蓋，放置起碼30分鐘的自動保溫，然後才將蓋打開。取出鮑魚，用針刺過鮑魚中心，如果很容易刺入，則表示鮑魚已經燜好。

⑧ 將鮑魚晾在架上，讓它與空氣接觸起碼2小時，這是做溏化程式。完成後，如果不是即食，將鮑魚用食物保鮮袋裝好，放入雪櫃凍格貯存。

⑨ 將高速煲內的材料及汁取出，用大網篩過濾出鮑汁。另用一小鍋，文火收鮑汁至濃稠但汁不要收得太少，要有足夠鮑汁留待將鮑魚重蒸時，汁要蓋過鮑魚面，必須緊記不要加入鹽。待汁涼卻後，用食物保鮮盒裝好放雪櫃凍格保存。

到想吃溏心鮑魚的時候……

① 把鮑魚預早取出室溫解凍，放入碗內，翻熱鮑汁，淋在鮑魚上，汁要蓋過鮑魚面，蒸15分鐘。

② 把鮑魚取出，盛在碟中。

③ 另用乾淨鍋把鮑汁煮沸，用1湯匙水加1茶匙生粉拌勻埋芡，試味，如果覺味道不夠，可以加入適量火腿汁或蠔油調味。

④ 在鮑魚淋上調好味的鮑汁，即可上桌。

火腿汁做法

　　金華火腿片（或雲南火腿）5克，加入白糖1茶匙、紹興酒1湯匙、水2湯匙，醃10分鐘，然後用大火蒸約15分鐘，即成火腿汁。火腿汁是作調味之用，而火腿片可以用於其他菜式中。

青咖喱牛面頰

愛吃牛的人不容錯過

材料

牛面頰（約400克）1塊	紅蘿蔔200克
馬鈴薯200克	泰國圓茄子4個
乾葱（切片）4粒	蒜頭（切片）3瓣
泰國青咖喱醬3湯匙	鹽1/2茶匙
糖2湯匙	椰漿100毫升

份量
2-3人份

準備時間
15分鐘

烹調時間
35分鐘

　　牛是吃草的動物，不停的咀嚼，使牛面頰的肌肉肉質堅實，不帶任何肥肉，有異於牛的其他部分。牛面頰不容易煮到很腍，優點是肉味特別濃，是愛吃牛的人刁鑽之選，只要烹調得宜，是一道非常美味的菜式。現代廚房，用電子高速煲來做紅燒牛面頰，烹調時間只需30分鐘，方便快捷，肉質軟滑入味，可配米飯、麵條、薯蓉或麵包。

　　牛肉的分割是根據各地的飲食文化而有所不同，名稱也不一樣。牛面頰在超市出售，一般沒有標明，而街市賣牛肉的檔口，牛面頰是放在冷櫃裏。畢竟一頭牛只有兩塊面頰，加起來的重量也不夠一公斤，可以預先向相熟的牛肉檔訂購，可以買一塊或兩塊，但通常不會分切出售。牛面頰買多了可以放在冰格裏長時間保存，用的時候才解凍，肉質不會受影響。

做法

① 把牛面頰兩面的筋膜起淨,洗淨,瀝乾水分。

② 紅蘿蔔和馬鈴薯削皮後切件,泰國圓茄子切去蒂,每個切成兩塊。

③ 加入1湯匙油在電子高速煲的鍋內,設置「輔助烹飪」10分鐘,按「Start」開始,放入乾葱片和蒜片炒香,再放入青咖喱、鹽和糖拌勻成醬汁。

④ 放入牛面頰、紅蘿蔔、馬鈴薯和圓茄子,與醬汁拌勻。

⑤ 把牛面頰放在煲底,上面是紅蘿蔔、馬鈴薯和圓茄子。

⑥ 在煲邊加入2湯匙清水。

⑦ 加蓋,設置「High」高壓20分鐘,按「Start」開始。

⑧ 待洩壓後,打開高速煲,先把牛面頰取出切件,再放回煲內。

⑨ 加入椰漿,把所有材料拌勻即成。

小碗薄切梅菜扣肉

細細份、輕輕的、靜靜食、好好味

梅菜扣肉，是傳統客家名菜，我們從小就認識它，算得上是香港人的集體回憶之一。梅菜扣肉雖然十分美味，但往往給人一種肥膩的感覺，似乎想一下都有罪惡感！而且自己做要燉上幾個小時，很多人都會宣佈放棄。近年在餐館吃到的梅菜扣肉，做法越來越粗糙，很少有滿意的，加上份量不少，往往吃不完，思前想後，還是不要打包了！

我們為您設計的這道「小碗薄切梅菜扣肉」，就解決了所有的問題。優點是：份量小，一餐可以吃完；肉片切得輕薄，入口不會感覺油膩，但依然非常入味；做法快捷簡單，把材料放入電子高速煲，設置開掣後，不用看顧，可以輕輕鬆鬆上班去，回家只要煮上米飯，開餐可也！

小貼士

如果用的是帶葉的梅菜而不是梅菜芯，一定要把每一片葉子張開沖洗裏面的鹽和沙。

份量
2人份

準備時間
20分鐘

加熱時間
40分鐘

材料

五花腩150克

老抽1/4湯匙

紹興酒1/2湯匙

甜梅菜芯75克

糖1湯匙

薑汁1/2湯匙

做法

① 刮淨五花腩皮上的毛,用清水沖洗,再氽水15分鐘,用清水沖至涼。

② 梅菜芯用清水泡浸5分鐘,在水喉下沖洗乾淨,擠乾水分。

③ 用刀切除梅菜頭較硬的部分,再把梅菜剁碎。

④ 用糖、酒和薑汁把梅菜拌勻,再拌入2湯匙油。

⑤ 把五花腩切成2毫米厚的肉片,加老抽拌勻。

⑥ 取一個500毫升容量的小碗,先把肉片平鋪在碗底和碗邊,再放上梅菜,用手把梅菜輕輕壓平。

⑦ 在電子高速煲內放入蒸格,加入125毫升清水,在蒸格上放上盛梅菜豬肉的小碗。

⑧ 蓋上煲蓋,設置「High」高壓40分鐘,再按「Start」開始。

⑨ 高速煲減壓後,打開,取出小碗,先用小碟按住梅菜,漤出碗內的汁,再把梅菜扣肉反扣在另一個碟子上,然後把汁倒回在肉片上即可享用。

清湯獅子頭

湯清澈有鮮味，肉質入口即溶。

一般餐館做的，絕大多數是紅燒獅子頭，把肉丸預早做好炸熟，有食客點菜時就下鍋再炸及加醬汁紅燒。揚州清湯獅子頭，味道和口感遠勝紅燒獅子頭，但很少餐館會做。要做到肉丸鬆而不散開，入口即溶，湯要清澈而鮮味，對餐館來說，不及紅燒獅子頭那麼容易做，而且不易做得好。

但是，只要您吃過清湯獅子頭，您可能以後都不再想吃餐館的紅燒獅子頭。做得好的揚州清湯獅子頭，要燉起碼兩小時，肉才會夠軟糯。我建議用電子高速煲，剛可放入品字形排列的三個小湯盅（甜品盅），只要用高壓加熱20分鐘，做出來的清湯獅子頭，口感和味道的效果，與在鍋中燉兩小時一樣。

份量
3小盅

準備時間
15分鐘

烹調時間
20分鐘

材料

絞脢頭肉150克　　　　　　肥豬肉100克

鮮蝦50克　　　　　　　　麥片1/2湯匙

娃娃菜1棵　　　　　　　　薑汁1/2湯匙

鹽1/2茶匙　　　　　　　　糖1/8茶匙

生粉2茶匙　　　　　　　　胡椒粉少許

雞湯適量

小貼士

① 可用普通鍋的做法，把盛丸子的小盅用鋁箔紙密
封，放在蒸爐中，隔水燉2小時即成。

② 要做到入口即溶，肥豬肉不可缺，切肥豬肉要先放
在冰格中凍至半硬，才容易切成小粒，不要用刀
剁，否則湯會有浮油。

③ 加入麥片可增加黏度，但不要把肉「撻」至起膠，
這肉丸會「彈牙」，但不會入口即溶。

做法

① 肥豬肉切成條，然後再切成半粒青豆大小。

② 鮮蝦去殼，剁成蝦茸。

③ 麥片用手捏成粉狀。

④ 娃娃菜剝開菜葉，洗淨。

⑤ 除娃娃菜和雞湯外，把所有材料放在大碗裏，用手捏到完全混合，把混合好的材料做成三個大丸子。

⑥ 把娃娃菜分別放在3個白瓷器或不銹鋼的小盅裏，再放入肉丸，加雞湯至完全覆蓋肉丸表面。

⑦ 在電子高速鍋中放下蒸格，加125毫升水，放入三個小盅品字形排好，不用加盅蓋或封紙。

⑧ 蓋上高速煲，設置「High」高壓20分鐘，按「Start」開始。完成洩壓後即可上桌。

柱侯蘿蔔牛筋腩

只需要花35分鐘，
做出一鍋香噴噴原汁原味的牛筋腩

　　燜牛筋腩是很多香港人的至愛，牛腩燜得入味而軟腍，牛筋香滑而有咬口，最令人黯然銷魂的是那盡吸肉味的白蘿蔔，往往是反客為主，最快被搶得清光。

　　蘿蔔牛筋腩要燜得好，要花好些時間，牛腩、牛筋、白蘿蔔要分三段時間來煮，需時起碼兩小時，在我們的《最愛香港菜1》一書中有詳細介紹。現在介紹的是用電子高速煲來做，快捷方便，由於這道菜用電子高速煲烹調不用加任何水分，湯汁的來源是白蘿蔔和牛腩，所以保證原汁原味。這道菜配米飯，或來一大碗麵，美味而滿足！

材料

牛腩（坑腩）600克

牛筋300克

白蘿蔔600克

調味料

柱侯醬3湯匙　　八角2粒

花椒1湯匙　　生粉1湯匙

薑片30克　　紹興酒2湯匙

蠔油2湯匙　　糖1湯匙

羅漢果1片

做法

① 燒沸大煲清水，把洗淨的牛腩、牛筋放入煲中汆水5分鐘，撈出，用清水沖洗。

② 牛腩切成約4厘米的方塊，牛筋切成6厘米長段。

③ 白蘿蔔去皮，滾刀切成大塊。

④ 花椒、八角、羅漢果放在小香料袋中，紮好備用。

⑤ 把生粉拌入牛腩中，用手抓勻。

⑥ 把牛腩、牛筋、蘿蔔、薑片和柱侯醬等所有調味料拌勻，放進電子高速煲內，香料袋放在煲底。

⑦ 在高速煲設置「High」高壓20分鐘，再按「Start」開始。

⑧ 待洩壓後，打開煲蓋，把牛腩、牛筋、白蘿蔔連湯汁盛出即可享用。

份量
4人份

準備時間
10分鐘

烹調時間
25分鐘

柱侯蘿蔔牛筋腩

小貼士

用電子高速煲燜煮，並不需要加入水分，因為材料在高壓下會自然出水，若加了水，湯汁就會太多，而味道就淡了。

醬料介紹

柱侯醬是廣東佛山的名產，已有百多年歷史，由當時一名叫梁柱侯的牛腩牛雜小販發明。柱侯醬用黃豆、鹽、麵粉和水經傳統醱酵抽取頭抽後的豆渣，加入果皮、蒜頭、芝麻醬及糖等原料煮製而成。柱侯醬味道香濃，入口醇厚甘滑，是風味絕佳之粵式醬料，適用於燜製牛腩、雞、鵝、鴨及肉類，風味獨特。柱侯醬在各大超市有售。

柱侯醬

豬腳薑醋二人前

不需要勞師動眾，想吃就很快有得吃

很多人喜歡吃豬腳薑醋，根據清代汪昂撰的中醫古籍《本草備要》，醋「酸溫。散瘀解毒，下氣消食。開胃氣，散水氣。治心腹血氣病，產後血暈，癥結痰癖，疸黃癰腫。口舌生瘡，損傷積血，穀魚肉菜薑諸蟲毒」。

除了醋的功用外，薑則具有祛風、驅寒、暖胃的功效，豬手或豬腳含豐富的骨膠原和鈣，雞蛋則提供營養和蛋白質。現代物資豐富，坐月子補身的方法很多，不一定要吃薑醋，但由於豬腳薑醋酸甜可口，成為很多人喜愛的食物，在酒樓上也成為一道受歡迎的點心。

煮豬腳薑醋，再也不用動不動煮一大鍋，使用電子高速煲，只需一個小時就煮好。

材料

雞蛋2隻　　　　　　　　　　豬蹄1隻

薑80克　　　　　　　　　　鹽（醃薑用）1茶匙

添丁甜醋300毫升（實際份量要以能浸過豬蹄和薑為準）

份量
2人份

準備時間
10分鐘

醃薑時間
30分鐘

加熱時間
40分鐘

做法

① 先準備一小煲清水，放入室溫的雞蛋，轉慢火把水燒至微滾開始計算煮約3分半鐘，取出，用冷水沖至涼卻，即成溏心雞蛋，剝殼後放在甜醋中浸泡至上顏色。

② 豬蹄請肉店把豬毛刮洗乾淨，斬成六件，汆水10分鐘，再用清水把豬蹄沖至涼卻。

③ 薑刨皮，切成粗條，拍裂。

④ 把薑用鹽醃30分鐘，再用清水把鹽洗淨。

⑤ 把蒸格放在高速煲內，加125毫升水。

⑥ 在大碗中加入薑和甜醋，把整個碗放在高速煲內，蓋上煲蓋，設置「High」高壓15分鐘，按「Start」啟動。

⑦ 當高速煲完成洩壓時，掀起煲蓋，把豬蹄放入容器內，讓醋浸過豬蹄。

⑧ 蓋上煲蓋，設置「High」高壓20分鐘。按「Start」啟動，洩壓後即成。

⑨ 吃時放上溏心雞蛋，即可享用。

小貼士

① 用電子高速煲煮豬腳薑醋用的大碗，口徑不要太大，否則要加入很多醋才可以讓醋浸過豬蹄，而且煮好後取出時會比較困難。

② 雞蛋如果放入與豬蹄同煮就會煮成老蛋，所以只能用醋浸泡入味。

番茄洋蔥燴牛脷

享受濃濃的地中海風情

博洛尼亞(Bologna) 是在意大利北部的一個古老的城市，是一個文化、美食、音樂和學術的中心，其中博洛尼亞大學建校近一千年，是世界上最古老的大學之一。博洛尼亞的一個傳統的菜式「肉醬意粉」(Spaghetti Bolognese)，和肉腸 (Bologna sausage) 也是聞名世界。

Spaghetti Bolognese的做法是用番茄或番茄醬、洋葱、碎牛肉（或豬肉），加上混合香料，做成肉醬汁，淋在意大利粉上，灑上帕馬臣芝士粉，就是風行世界的肉醬意粉，我們喜歡以這美味的醬料用來燴牛脹。

吃過牛脹的人都會愛上它那特別的口感，很多人都以為牛脹很難處理，其實做法很簡單，基本上就是加水耐心地煮至脹，並無特別技巧。用電子高速煲來煮牛脹，煮的時間就節省很多，保證牛脹一定會煮得夠軟脹，而且由於電子高速煲煮食不會失去水分，所以醬汁是原汁原味。

份量
4-6人份

準備時間
20分鐘

烹調時間
25分鐘

番茄洋葱燴牛脹

材料

牛脷（約1.4千克）1條　　　　番茄300克

洋蔥300克　　　　　　　　紅蘿蔔300克

混合香草1湯匙

（意大利或普羅旺斯混合香草）

番茄膏（170克）1罐　　　　鹽1.5湯匙

糖2湯匙

做法

① 用利刀把牛脷上白色的厚皮切掉（在買牛脷時可請店家代勞），洗淨。在舌頭和舌根處切斷。

② 把舌頭切成1厘米厚片，舌根從中間切開成兩邊，再切成1.5厘米厚片，放入清水中浸洗10分鐘，取出瀝水。

③ 番茄洗淨，在底部用刀剕十字，用沸水燙2分鐘，把皮剝掉，再切成八大塊。

④ 洋蔥剝皮，切成八大塊。

⑤ 紅蘿蔔削皮，切成1.5厘米厚片。

⑥ 把牛脷放在電子高速煲底部，再放上洋蔥和紅蘿蔔。

⑦ 把番茄膏和番茄、鹽、糖、香草拌勻，放在洋蔥、紅蘿蔔和牛脷上。

⑧ 蓋上高速煲的蓋，設置「High」高壓20分鐘，按「Start」，煮好即成。

小貼士

① 牛舌頭和舌根的肉質不一樣，舌頭較硬，舌根較軟，所以切的厚度也應該不一樣。

② 由於番茄和洋蔥會出很多水分，所以用高速煲來煮，完全不用加入水分。

昆布紫菜排骨湯

調理失眠和肥胖的美味湯水

昆布即海帶，性寒，有清熱、安神、降血壓的功效，是一種非常健康的食物。有人說，十個女人九個都有不同程度的水腫，常吃昆布，可幫助消腫利水，調理肥胖症及腳氣浮腫。

很多都市人因各種原因，失眠夢多睡不好，其中不少人失眠的原因，是中醫所說的痰熱偏盛的失眠，伴有心煩、口苦、頭重等症狀，而吃這道昆布紫菜排骨湯就最適合了！

做法

① 把昆布用清水泡30分鐘，去沙洗淨，剪成段。

② 排骨洗淨，瀝乾水分。

③ 把排骨和昆布放進電子高速煲內，加薑和750毫升水。

④ 蓋上煲蓋，設置「High」高壓20分鐘，按下「Start」啟動程序。

⑤ 洩壓完成後，打開煲蓋，趁熱放入紫菜和鹽拌勻即可上桌。

材料

昆布（乾）10克　　紫菜5克

排骨300克　　薑片5克

鹽1/2茶匙

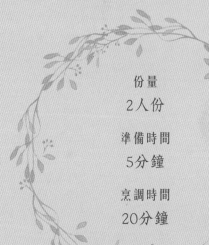

份量
2人份

準備時間
5分鐘

烹調時間
20分鐘

小貼士

① 昆布（海帶）可在超市及街市買到。

② 建議買壽司用紫菜，不用煮，用熱湯泡軟即可。

③ 或用普通鍋，煮沸後轉小火煲兩小時即可。

鮮人參雞湯

延緩衰老，補血益氣睡眠好！

份量
2-3人份

準備時間
30分鐘

浸泡時間
2小時

烹調時間
40分鐘

材料

土雞或烏雞（淨重約1千克）1隻

鮮高麗人參（約30至35克）1條

糯米120毫升　　　　　　　紅棗3粒

蒜頭4瓣　　　　　　　　　栗子肉8粒

白果肉12粒　　　　　　　鹽（浸糯米用）1/2茶匙

鹽（煮湯用）1茶匙　　　　水1公升

人參雞湯是到韓國旅行必吃的食物，大長今更讓我們見識了韓式養生！韓國人深受儒家思想薰陶，推崇「藥食同源」的飲食方法，烹調方法自然純樸；在人參雞湯中加入栗子、白果和糯米，中和了雞湯的燥熱。

在家做一鍋暖胃滋補的鮮人參雞湯，除了上班前預早浸糯米之外，回家用電子高速煲1小時搞掂。有湯有餸，對自己好些，對家人好些，優雅體貼，柔情滿溢！

① 糯米加入鹽拌勻，用與糯米等量的清水浸約2小時，用清水洗淨，瀝去水分。

② 把雞腔內的肺和腎清除，撕去脂肪，切去雞頸、雞頭不要，洗淨瀝乾，雞爪在雞腿關節下1厘米砍斷留用。

③ 鮮人參用軟刷子刷去表面的泥土，洗淨，再切成兩段。

④ 紅棗去核，撕去掉紅棗皮。蒜瓣原粒去衣。

⑤ 把糯米、蒜瓣和鮮人參塞進雞腔，用線把尾巴開口處縫上。

⑥ 用棉繩把兩條雞腿在關節處縛在一起，使雞腿合攏，兩隻雞翅膀翻向後貼着雞背。

⑦ 把雞放在高速煲內，再放入雞爪、白果肉、栗子肉、紅棗肉和鹽，加入清水。

⑧ 設置「High」高壓40分鐘，再按「Start」開始。

⑨ 待自動洩壓後，取出雞和湯放在大碗中，拆去棉繩和線即可享用。

小貼士

① 如果採用其他湯煲，把浸過的糯米、人參及其他配料放入切去頭頸的雞腔中，用線縫好。煮沸3000毫升水，把雞放入，加蓋煮沸，轉小火煮3小時即成。

② 鮮人參在韓國人食品超市有售，買回家後不要清洗，包好放冰箱中可貯存數天。

③ 把紅棗皮去掉，可使湯的顏色較為清澈。

羅宋湯

用電子高速煲，做一鍋健康美味的風味湯

材料

豬肉（或牛肉）300克	蒜頭4瓣
紅菜頭1/2個	馬鈴薯150克
紅蘿蔔150克	椰菜（約300克）1/2個
洋葱150克	牛油1湯匙
番茄膏（170克）1罐	香葉1片
鹽1茶匙	酸忌廉

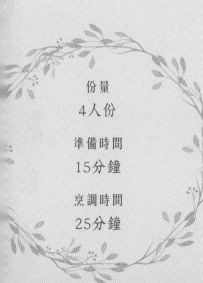

份量
4人份

準備時間
15分鐘

烹調時間
25分鐘

很多人以為羅宋湯是源於俄羅斯，可能是因為「羅宋」二字。在二十世紀中期，香港開設了好幾家俄羅斯菜的餐廳，餐牌上都一定有羅宋湯，其實羅宋湯原創於烏克蘭，後來傳到俄羅斯和其他東歐國家如白俄羅斯、波蘭、立陶宛等地，正確的英文名稱應該是 borscht，即紅菜頭湯。

羅宋湯在歐洲，可以熱吃或冷吃，是一道冬夏皆宜的家常菜式，味道微酸，傳統上會加入發酵過的紅菜頭汁、番茄或醋。羅宋湯所用的材料很隨意，可以用豬、牛、雞、香腸、紅菜頭、洋葱、馬鈴薯、紅蘿蔔、番茄、椰菜或其他蔬菜，而紅菜頭是羅宋湯不可缺少的材料。吃時加入酸忌廉，更是羅宋湯的原裝正版。

做法

① 肉洗淨，切成2厘米的肉丁，和蒜頭一同放進電子高速煲，加1公升水，設置「High」高壓10分鐘，煮成肉湯。

② 在煮肉湯時，把紅菜頭、馬鈴薯、紅蘿蔔切1.5厘米方粒。

③ 椰菜和洋葱切成粗絲。

④ 用牛油把洋葱炒香，和椰菜、紅菜頭、馬鈴薯、紅蘿蔔、番茄膏、香葉和鹽一同放進肉湯內拌勻。

⑤ 再設置「High」高壓10分鐘，待洩壓後即成。

⑥ 吃時伴以酸忌廉。

L'ÉPICE LA PLUS DÉLICIEUSE

花膠響螺頭燉雞湯

半小時做出滋補明目的養顏靚湯

響螺頭是廣東人的滋潤湯水最佳材料，有滋陰養顏、明目、補腎養肝、止夜尿等功效。香港市場上的主要品種有美國急凍響螺頭、澳洲白螺頭、青島螺頭、南美螺頭，紅螺頭和智利的螺片。乾貨響螺頭可用冷水浸一晚，再用薑葱出水即可用來煲湯。急凍響螺頭則比較簡單，解凍即可用，而且口感柔軟，味道鮮甜。一般用明火煲螺頭湯，要煲3至4小時，如果用電子高速煲的話，大約半小時即成。連湯帶渣吃，鮮美無比，最適合工作忙碌的上班一族。

材料

花膠（未發重量）60克

急凍響螺頭（約250克）2件

雞（約500克）1/2隻

豬腱150克

薑片10克

鹽1/2茶匙

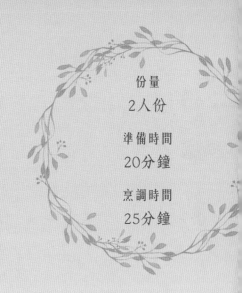

份量
2人份

準備時間
20分鐘

烹調時間
25分鐘

做法

① 花膠預先發好（見本書第62頁）。

② 把急凍響螺頭解凍洗淨，去掉腸子內臟，切除響螺頭上的萼蓋，再切成1厘米厚的螺塊。

③ 把雞去皮，切去雞頭、雞尾及撕去雞肺、腎、筋膜，洗淨。

④ 豬腱洗淨，切成小塊。

⑤ 把螺片、雞、豬腱和薑片放進電子高速煲，加鹽和750毫升水，蓋上煲蓋。

⑥ 高速煲設置「High」高壓20分鐘，按下「Start」。

⑦ 洩壓完成後打開，放入花膠，再設置「Low」低壓5分鐘，洩壓後即成。

小貼士

如果用普通湯鍋，按上述①至④做好，放入湯鍋，大火煮沸，再用小火煮3小時，然後放入發好的花膠，再煮30分鐘即成。

香茅薑茶

低卡路里低糖的清涼飲品

　　減肥人士要控制飲食，汽水、奶茶、樽裝飲料全部要減少飲用。對自己好一些，輕輕鬆鬆自製一杯香茅薑茶，在炎炎夏日，倍感清新舒暢，醒腦提神，方便快捷。香茅在超市及街市均有售，用濕布或保鮮紙包好，放在冰箱中可貯存數日。

材料

新鮮香茅3枝

薑（切片）40克

冰糖50克

份量
1公升

做法

將香茅用水沖洗後，切段，用刀背拍裂，與薑片同放入電子高速煲中，加入1公升水及冰糖，設置「High」高壓5分鐘，再按「Start」開始，待洩壓後稍涼倒出，用網篩隔去渣即可熱飲或放入雪櫃成凍飲。

小貼士

如果用普通湯鍋，將香茅用水沖洗後，切段，用刀背拍裂，與薑片同放入鍋中，加入1.5公升水，大火煮沸後轉小火煮1小時，加入碎冰糖，煮至糖粒溶化，倒出，用漏網隔去渣即可熱飲或凍飲。

南北杏雪梨水

生津潤肺，令你人靚聲甜

材料

南北杏（約1兩）40克（南北杏比例 4：1）

雪梨2個

冰糖40克

份量
1公升

做法

① 南北杏用水洗淨瀝乾。

② 雪梨不削皮，用水洗淨，切成四邊，切去梨芯。

③ 把雪梨與南北杏和冰糖一起放入電子高速煲中，加入1公升水，
設置「High」高壓15分鐘，再按「Start」開始。

④ 待洩壓後，稍涼可以倒出飲用。

小貼士

如果用一般燉盅，將南北杏用水洗淨；雪梨不削皮，用水洗
淨，切成四邊，切去梨芯，與南北杏和冰糖一起放入燉盅內，
加入1公升水，密封，隔水燉1小時即成。

越能幹的人，工作就越忙碌，在辦公室裏一坐就是九小時或以
上，有些行業的工作時間更長，早出半夜歸。由於長時間困在大樓的
空調環境中，皮膚因缺水而慢慢變得乾燥，加上肺部運動量少，就較
容易有咳嗽甚至失聲等問題。南北杏雪梨水，潤喉潤肺，滋養皮膚，
潤腸通便，越飲越靚！

黑木耳紅棗茶

預防血脂、血糖、血壓三高的保健茶水

材料

白背木耳10克

去核開邊紅棗8粒

薑3片

冰糖40克

份量
1公升

做法

白背木耳用水浸1小時，沖洗乾淨後，撕成小塊，和紅棗、薑片、冰糖和1.25公升水，放入電子高速煲中，設置「High」高壓15分鐘，再按「Start」開始，待放壓後稍涼可以倒出，用網篩隔去渣即可飲用。

小貼士

如果用普通湯鍋明火烹調，將白背木耳用水浸1小時，沖洗後放入鍋中，加入去核紅棗、冰糖及薑片，倒入1.5公升水，大火煮沸後轉小火煮2小時，倒出，用漏網隔去渣即可飲用。

　　現代人最怕身體有三高，黑木耳能降低人體血脂和膽固醇，延緩血管硬化，減少心臟病及腦中風的風險。黑木耳更是糖尿病者最有益處的食物，能增加及改善胰島素的分泌，以及減少人體血糖的波動，常飲能令您遠離三高，延年益壽。

糯米蒸排骨

用新一代的磁應電飯煲，
煮糯米不用預先浸泡。

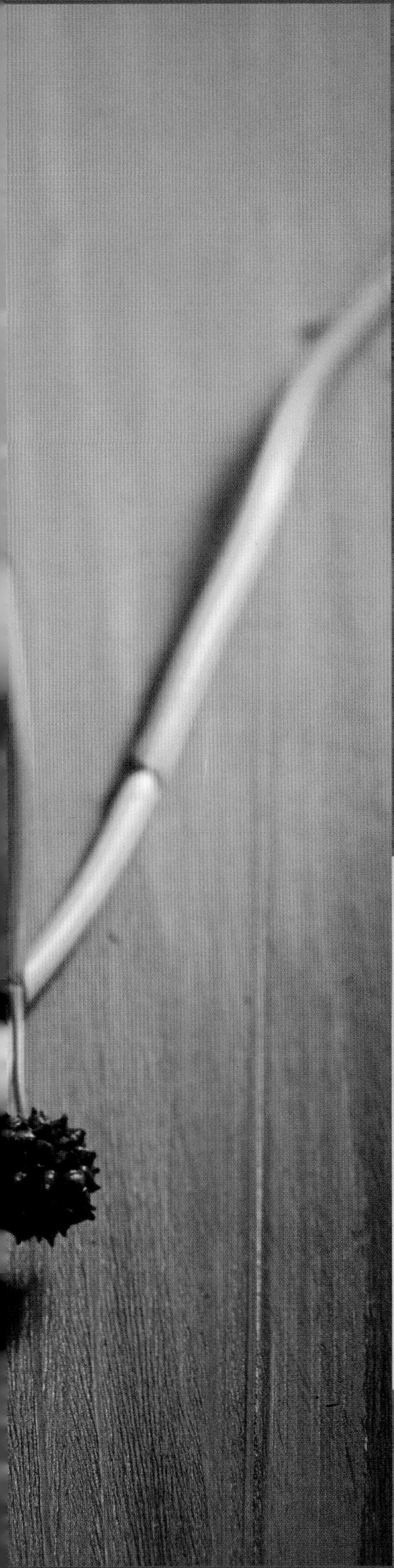

　　新一代的磁應電飯煲，特點是有浸米的功能，也因為有這功能，煮飯的時間比普通電飯煲長，而米粒經過浸泡，會煮得全透，煮出來的飯粒粒都是飯芯，米香味更能散發，口感也較佳，這就是智慧型磁應電飯煲的價值所在。

　　我們煮或蒸糯米飯時，一般都要先把糯米預早用水浸泡，這樣才可以減少煮的時間，而糯米飯口感會更軟。用磁應電飯煲來煮糯米飯，就可以直接下糯米，不用預早浸泡。

　　糯米蒸排骨是一道順德菜，曾在我們的《巧手精工順德菜》一書介紹過，今次是用電飯煲來做，烹調簡易，更適合新手下廚的您！

材料

糯米250毫升

鹽1/2 茶匙

葱1條

生抽1湯匙

糖1/2茶匙

排骨（斬件）200克

蒜頭2瓣

南乳（小塊）1塊

料酒1茶匙

生粉1/2茶匙

份量
2至3人份

準備時間
10分鐘

醃製時間
15分鐘

烹調時間
1小時

做法

① 糯米洗淨，瀝乾水分。

② 蒜頭剁蓉，葱切葱花，南乳壓爛，加2湯匙水拌成醬汁。

③ 排骨洗淨後，用醬汁、生抽、料酒、糖和生粉醃15分鐘。

④ 把糯米放入電飯煲內，加入鹽和水，水量要比正常水量少40毫升，選擇「Delicious模式」設置。

⑤ 見飯煲中的水沸時，放入排骨排好。

⑥ 飯煮好後，不要馬上開蓋，再焗5至10分鐘，取出排骨，把飯拌勻即成。

小貼士

① 煮糯米所需的水比較煮白米的水要少，否則糯米飯會太爛，而且排骨也會溢出一點水，所以水量要比正常水量少40毫升。

② 糯米飯煮好後要再焗幾分鐘，是要確保排骨全熟。

三文魚蘆筍野菌飯

用電飯煲煮出美妙的健康組合

都市人工作繁忙，往往不能準時下班，晚上匆匆回家，雖然肚子餓，卻更不想再勞累煮食。介紹的這味三文魚蘆筍野菌飯，就最適合忙碌人士。

三文魚配合蘆筍和野菌，營養均衡，少油而清淡，是注重健康的人士最佳的選擇。而且，三樣材料都可以隨時在超市買得到，十分方便。乾野菌可增加飯的香味，我們習慣用的是羊肚菌乾，香味很濃，雖然價格較貴，不過用量很少。也可以用浸透冬菇切粒代替，如果買新鮮菇類例如秀珍菇、蘑菇，購買方便，但菌味較淡。再簡單一點，也可以省去野菌或菇類。

這個飯煮好後，也可以不必煩惱如何把三文魚整片取出，而是挾出魚骨，用飯勺搗爛魚肉，拌入香飯中，完成一道美妙的懶人料理！

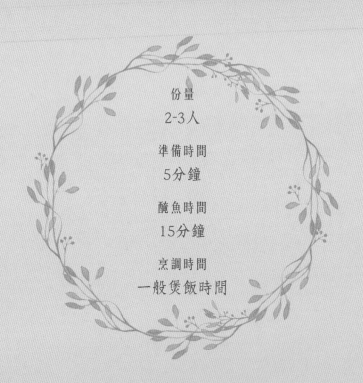

份量
2-3人

準備時間
5分鐘

醃魚時間
15分鐘

烹調時間
一般煲飯時間

材料

三文魚約160克　　　　味醂2茶匙

生抽2茶匙　　　　　　蘆筍4條

白米240毫升　　　　　鹽1/2茶匙

橄欖油1湯匙　　　　　羊肚菌乾8粒

清雞湯或水適量

做法

① 三文魚放在保鮮密實袋中，加入味醂和生抽，封口，用手搓勻袋中的魚和醃汁，醃15分鐘。

② 蘆筍切去淺色筍頭，洗淨。

③ 白米略為沖洗過，瀝乾水分。

④ 羊肚菌乾用60毫升清水泡10分鐘，剪開洗淨，取出，泡菌用的水過濾後留用。

⑤ 把白米、鹽、橄欖油放入電飯煲，加入泡菌的水和適量清雞湯（或水），按下「Start」煮飯。

⑥ 見飯水煮沸時，放入泡過的羊肚菌，繼續煮至米飯快將收水，即約最後5至10分鐘。

⑦ 把三文魚放入，蘆筍排在魚旁，煮至飯熟。

⑧ 不開蓋，再自動保溫焗5分鐘即可進食。

小貼士

把米、鹽、橄欖油放入電飯煲後，加入泡野菌水和適量清雞湯或水，加起來的水量，等於平日煮飯的水量，由於每家人吃的白米和用的電飯煲都不同，水量要自行調整。

簡易海南雞飯

教你用電飯煲一次過煮出香噴噴的美食

很多人都喜歡吃海南雞飯，但先要蒸白切雞，又要留雞汁，然後再弄那鍋油飯，步驟不少，很多人會怕麻煩。有沒有想過可以用電飯煲一次過煮出海南雞飯？相信這是很多上班一族的夢想吧！

份量
2至3人份

準備時間
15分鐘

醃製時間
30分鐘

烹調時間
一般煮飯時間加10分鐘

材料

光雞（約700克）1/2隻	白米320毫升
鹽2茶匙	薑汁1湯匙
紹興酒1/2湯匙	蒜頭（剁蓉）1/2個
香茅1根	椰漿100毫升

小貼士

① 用電飯煲一次過煮海南雞飯，米量不能太少，否則雞可能不夠時間焗熟，可再加長焗雞的時間至10分鐘。

② 320毫升的米，大約是電飯煲米杯的兩杯米。

③ 煮雞飯的時候，雞汁會流到飯中，所以要少放一些水。水量要視乎個人對米飯的喜好，以及按不同的白米而定。

④ 這個方法做海南雞飯，適用於可以中途打開蓋的磁應電飯煲或西施電飯煲，不適用於快煮的普通電飯煲。

做法

① 把雞洗淨，用1.5茶匙鹽、薑汁和酒醃30分鐘，中途把雞翻轉一次。

② 白米淘淨，瀝乾水分。

③ 用2湯匙油，慢火把蒜蓉爆香，加入白米和1/2茶匙鹽炒勻，放入電飯煲中。

④ 香茅撕去外皮，拍扁，洗淨，切成4段後，放在米上。

⑤ 加入水。水量要比平常用的飯水少約40至50毫升。在電飯煲選擇「Delicious模式」慢煮。

⑥ 當米水煮沸後，放入雞，皮朝上，醃雞的汁水淋在雞上，蓋好，繼續煮。

⑦ 當飯煮好跳掣後，先不要開蓋，讓雞和飯再焗5至8分鐘。

⑧ 開蓋取出雞和香茅，在飯中倒入椰漿拌勻。蓋上煲蓋再保溫焗5分鐘即可。

⑨ 待雞放涼後便可斬件，盛飯一同享用。

上海菜飯

有菜有肉又有飯，用電飯煲搞定！

　　由清朝末年開始，上海成為當時中國重要的對外通商口岸之一，碼頭和倉庫的貨運非常忙碌，附近農村有很多窮人跑到這些碼頭和貨倉當苦力，這可能是中國第一代的入城民工。這些民工在上海賺取血汗錢，生活十分節儉，碼頭附近有些小販，就用最廉價的上海青（小棠菜）加在有鹽的飯中同煮，賣給這些碼頭工人吃，成為當時稱為「苦力飯」的第一代菜飯。後來菜飯就在上海慢慢流行起來，還加入了上海鹹肉，從此，不再是「苦力飯」了。

　　上海菜飯可以配合菜肴一起作為主食，也可以簡簡單單地只吃一大碗菜飯，有肉有菜有飯。我父親堅持保留做上海菜飯要「有些少豬油」，說否則小棠菜缺油，飯味就不是那一回事了，而且肥肉炸油後的豬油渣非常可口，混在菜飯中偶然香脆一下，人間美食也！

份量
2至3人份

準備時間
10分鐘

浸泡時間
30分鐘

烹調時間
一般煲飯時間

材料

白米240毫升

上海鹹肉100克

肥豬肉30克

小棠菜100克

蒜蓉2湯匙

鹽1茶匙

① 白米洗乾淨後瀝乾。

② 上海鹹肉用清水浸半小時泡去部分鹹味，取出，切成1/2厘米厚片。

③ 肥豬肉切成小粒，備用。

④ 小棠菜洗淨，在開水裏迅速焯一下，取出，泡在冷開水至涼，瀝乾，切碎。

⑤ 在鑊中下1茶匙油，放進肥豬肉粒，用慢火炸出豬油後，豬油渣拿出留用。

⑥ 把蒜蓉用鑊中的豬油略炒到出味，加入白米同炒約半分鐘。

⑦ 把炒過的米放在電飯鍋裏，放進1茶匙鹽，按正常煮飯加水煮飯。

⑧ 當米飯開始收水時，在飯面放上鹹肉。

⑨ 飯煮好後，把切碎的小棠菜和飯拌勻，吃時撒上豬油渣即成。

小貼士

① 上海鹹肉可以在南貨店購買，最好買帶肥的，味道比較甘香。

② 小棠菜不要預先焯得太熟，否則拌入熱飯中會變黃。

皮蛋瘦肉粥

用聰明的磁應電飯煲煮出綿綿粥底

香港人喜歡吃粥，特別是那身體有少許不舒服的日子，就會想吃一碗皮蛋瘦肉粥。無論煲哪一款的粥，都先要煲一個白粥底，然後再加各式配料。一煲「靚」的廣東粥底，要求的是入口香滑，米粒完全化開，只是依稀能見，用羹勺盛起時，粥面微微凸起，就稱之為「綿」。我們陳家廚坊系列中，《回家吃飯》一書中，就有詳細介紹怎樣煲廣東明火白粥底。

新一代的智能磁應電飯煲，特點是有浸米的功能，而米粒經過浸泡才煮，米粒便煮得全透，就可以輕鬆地煮出「綿」的粥底，不用擔心沸瀉或煮焦。

份量
3至4碗

準備時間
10分鐘

醃肉時間
4小時

烹調時間
約2小時

材料

豬肉300克	生粉1茶匙
皮蛋2隻	白米80毫升
鹽1湯匙	

做法

① 豬肉洗淨，切成片，加1湯匙鹽醃約4小時，放入飯煲前用清水把鹽洗去，再用生粉把豬肉拌勻。

② 把每一隻皮蛋切成8塊，備用。

③ 白米洗淨瀝水，放入智能電飯煲中，加800毫升水，蓋好煲蓋，選擇白米和煲粥，設置時間為2小時，按啟動開始煲粥程式。

④ 在粥煮好前10分鐘，掀起煲蓋，放入豬肉和皮蛋，拌勻，蓋上煲蓋繼續煮到煲粥程式完成。

⑤ 程式完成後，不要立即打開，再焗10至15分鐘，打開蓋用勺拌勻即成。

陳家廚坊

小貼士

① 煲靚粥要用新米，舊米的黏性不夠，可用泰國絲苗加一半澳洲油粘米，如用東北大米、台灣米或日本米則更為適合。

② 由於豬肉已用鹽醃過，煮好的粥要試味後才決定是否再加鹽。

③ 煮完後再焗10分鐘，是讓米粒靜靜「綿化」的過程，用普通鍋煮粥，也一樣要繼續焗15分鐘，粥底才會「綿」。

五味手撕雞
Shredded Chicken Salad in Pungent Sauce

Serves *2 to 4* / Marinate *30* mins / Cooking time *30* mins

Ingredients
½ chicken, about 600 g
1 tsp salt
1 tsp white sesame seeds
40 g pickled ginger
5 pc pickled jiao bulb
½ pc cucumber

Sauce
1 tsp light soy sauce
1 tsp sugar
1 tsp Zhejiang vinegar
½ tsp chili oil
1 tsp hot mustard
1 tsp sesame oil

Method
1. Marinate chicken with salt for 30 minutes.
2. Bring to a boil a large pot of water, put in chicken, turn off heat and steep chicken in hot water for 15 minutes. Remove chicken to a colander to drain.
3. When chicken is cool enough, debone chicken and tear both meat and skin into shreds.
4. Roast sesame seeds in an unoiled pan.
5. Cut pickled ginger and jiao bulbs into fine shreds and mix in with chicken.
6. Rinse cucumber and cut into fine strips.
7. Mix soy sauce, sugar, vinegar, chili oil, hot mustard and sesame oil into a pungent sauce.
8 Mix chicken with pungent sauce and cucumber, and top with sesame seeds before serving.

Tips

1. Rice vinegar and sugar are used to pickle jiao bulbs.
2. Wasabi can be used in place of hot mustard.
3. Shredded cucumber should only be added to the chicken just prior to serving so that the pungent sauce will not be diluted by cucumber juice.

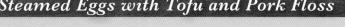

肉鬆豆腐蒸水蛋

Steamed Eggs with Tofu and Pork Floss

Serves *2 to 4* / Preparation *10* mins / Cooking time *10* mins

Ingredients

1 pc soft tofu, about 250 g
10 g preserved mustard green
2 pc egg
250 ml chicken broth
3 tbsp dried pork floss
½ tsp salt
2 stalk chopped green onion

Method

1. Cut tofu into 1.5 cm cubes and marinate with ¼ teaspoon of salt for 10 minutes. Drain.
2. Soak preserved mustard green for 10 minutes, and chop.
3. Beat eggs and filter through a sieve. Mix eggs with chicken broth and ¼ teaspoon of salt into an egg batter.
4. Put tofu into a plate, add egg batter and top with chopped preserved mustard green. Cover plate with microwave cling wrap and steam over high heat for 8 to 10 minutes or until eggs are fully cooked.
5. Top with dried pork floss and chopped spring onions.

Tips

1. Steaming eggs using induction heating or over fire requires 8 to 10 minutes after the water has come to a boil. If a steam oven is used, allow sufficient time for per-heating.
2. The time required to steam eggs depends very much on the shape of the plate. A deep plate will require 1 or 2 extra minutes.

碗蒸臘味蘿蔔糕
Turnip Pudding in a Bowl

Serves *3* / Preparation *10* mins / Cooking time *40* mins

Ingredients
600 g turnip
20 g dried shrimps
1.5 pc Guangdong cured sausage
3 tbsp rice flour
1.5 tbsp corn starch
½ tsp salt
⅛ tsp ground white pepper

Method
1. Peel and cut turnip into thick strands. Put turnip strands in a pot and simmer over low heat until they turn translucent. Set aside to cool.
2. Rinse and soak dried shrimps until soft. Chop.
3. Steam sausage until soft and cut into pea size bits.
4. Heat 1 tablespoon of oil and stir-fry shrimps and sausage.
5. Mix rice flour, corn starch and turnip strands, and add salt, ground white pepper, chopped shrimps and sausage. Mix thoroughly into a turnip paste.
6. Divide turnip into three portions and put into small bowls. Steam over high heat for 30 minutes.

Tips
1. The amount of juice in the turnip will vary depending on its freshness and some water may have to be added.
2. The ratio between rice flour and corn starch is 2:1. If a firmer consistency is desired, increase the amount of rice flour.

Serves *4* / Preparation *10* mins /
Marinate *45* mins / Cooking time *10* mins

Ingredients

300 g pork shoulder	1 tbsp salt
½ pc onion	½ pc green sweet pepper
½ pc red sweet pepper	1 small can pineapple
1 tbsp chopped garlic	½ tbsp light soy sauce
½ tsp cooking wine	½ tsp sugar
1 pc egg	4 tbsp gluten free flour

Ingredients for sweet and sour sauce

4 tbsp red Zhejiang vinegar   4 tbsp red sugar

Method

1. Cut pork into small chunks and marinate in 500 ml of water and 1 tablespoon of salt for 30 minutes. Drain.
2. Cut onion into large pieces, and sweet peppers and pineapple into smaller chunks. Do not use the syrup from the can.
3. Mix vinegar and red sugar in a bowl until the sugar is completely dissolved to become a sweet and sour sauce.
4. Marinate pork with garlic, soy sauce, wine and sugar for 15 minutes.
5. Beat egg and mix with pork.
6. Mix the pork with gluten free flour.
7. Heat 250 ml of oil over medium-low heat and deep fry pork until golden brown. Remove pork from oil.
8. Pour out oil leaving only 1 tablespoon in the wok, stir-fry onion until soft, add sweet and sour sauce, and sauté until sauce thickens. Put in sweet peppers and toss rapidly a few times. Add pork and sauté until the pork is fully coated with sweet and sour sauce. Stir in pineapple and transfer to plate.

Tips

1. Either fresh or canned pineapple or strawberries can be used.
2. The sweet and sour sauce when heated becomes a syrup and adheres to foods naturally without the use of a thickening agent. A well-made sweet and sour pork should have little sauce left on the plate.

班尼迪蛋
Eggs Benedict

Serves *2 to 4* / Cooking time *20* mins

Ingredients
4 pc egg
2 pc English muffin
4 pc ham
½ tsp salt
1 tbsp white vinegar

Hollandaise sauce ingredients
2 pc egg yolk
1 g salt
1 g sugar
½ tbsp lemon juice
100 g unsalted butter

Method
1. Slice English muffin into two halves, brush on butter and toast in a skillet or an oven. Pan fry sliced ham in a skillet.
2. Put toasted muffin halves on plates and top each with a slice of ham.
3. Crack an egg into a cup. If the yolk is broken, change the egg.
4. Put about 1.5 litre of water in a pot and bring to a boil. Add salt and vinegar, and reduce to low heat. Stir the water a few times with a long handle spoon to create a small whirlpool, and slide the egg following the direction of the water flow into the center of the whirlpool. Let the egg steep in hot water for 3 to 4 minutes until the desired firmness is reached. Remove the egg to a piece of kitchen paper towel to absorb excess water.
5. Heat the water in the pot again and repeat the above with the remaining eggs.
6. Place the cooked eggs on the ham and top with Hollandaise sauce.

Making the Hollandaise sauce
1. Melt butter and run it through a sieve to filter out the milk solids. Put clarified butter into a measuring cup.
2. Put egg yolks, salt, sugar and lemon juice into the tall cup that comes with the hand blender. Mix the ingredients with the hand blender and add heated butter gradually in small quantities to blend completely with the ingredients. Move the blender up and down while blending to obtain a smoother texture.
3. Pour the Hollandaise sauce into a bowl and cover the bowl to keep the sauce warm.

Tips

1. When making the Hollandaise sauce, add butter only when the previous batch of butter is completely blended with the other ingredients.
2. The acidity of the vinegar helps the egg whites to coagulate rapidly. Egg whites can easily be broken up if acidity is too low in the water.
3. Make only a small whirlpool in the water for the egg to slide in. Too large a whirlpool can cause the egg whites to break up.
4. Lower the cup as close to the water as possible when sliding the egg into the water. A cup with a handle works best. If the cup is held left handed, stir the water in a counter clockwise direction. If right hand is used to hold the cup, stir the water in a clockwise direction. This is to ensure the egg follows the flow of the water.
5. Eggs Benedict can also be toasted to a golden brown in an oven. However, care must be exercised to avoid over cooking the eggs.

韓式泡菜煎餅
Pancake with Pickled Vegetables, Korean Style

Serves *1 to 2* / Preparation *10* mins / Cooking time *5* mins

Ingredients

50 g Korean pickled cabbage
1 stalk green onion
⅛ tsp salt
1 pc egg

10 g carrot
50 g Korean pancake flour
⅛ tsp sugar
50 ml fresh water

Method

1. Cut pickled cabbage into thick strips.
2. Peel and shred carrot. Cut spring onion into thick strands.
3. Mix pancake flour together with salt, sugar, egg and water into a pancake mix.
4. Stir pickled cabbage, carrot and spring onion into the pancake mix.
5. Heat 1 tablespoon of oil in a flat non-stick skillet, put in pancake mix and pan fry over low heat to a light brown. Flip over pancake and lightly brown the other side. Transfer pancake to a plate.
6. Cut pancake to 8 pieces before serving.

豉汁豬頸肉蒸腸粉
Steamed Rice Roll with Pork Jowl in Black Bean Sauce

Serves *2 to 3* / Preparation *10* mins /
Marinate *10* mins / Cooking time *10* mins

Ingredients

300 g pork jowl
2 tbsp preserved black bean
1 tsp sugar
½ tsp cooking wine
1 pc red chili pepper

450 g rice roll
1 tbsp chopped garlic
1 tbsp light soy sauce
1 tsp corn starch

Method

1. Wash and chop preserved black beans, and mix together with garlic, sugar, soy sauce, wine and 2 tablespoons of water into a black bean sauce.
2. Rinse and slice pork jowl, add black bean sauce and marinate for 10 minutes.
3. Mix in corn starch and stir in 2 tablespoons of oil just prior to steaming.
4. Deseed chili pepper and cut into thin strips.
5. Cut rice rolls into 4 to 5 cm lengths, and distribute evenly on a plate.
6. Put marinated pork jowls together with the marinating sauce on the rice rolls and top with chili pepper. Steam over high heat for about 10 minutes.

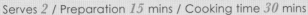

滑蛋蝦仁煎米粉

Scramble Eggs with Prawns over Rice Vermicelli

Serves *2* / Preparation *15* mins / Cooking time *30* mins

Ingredients
150 g dried rice vermicelli
300 g shrimp
4 to 5 egg
1 tsp salt
oil as needed

Method
1. Bring to a boil 1 litre of water in a pot, add rice vermicelli and ½ teaspoon of salt, cover pot and turn off the heat. Disperse vermicelli with chopsticks after 10 minutes, drain, and mix in 2 tablespoons of oil.
2. Blanch shrimps and drain.
3. Heat 2 tablespoons of oil over high heat in a non-stick skillet, put in vermicelli and spread it evenly over the skillet. Reduce to medium heat and brown the vermicelli. Flip over the vermicelli and brown the other side. Transfer to a large plate.
4. Separate egg whites from the yolks and beat egg whites until foamy. Blend in 1 teaspoon of oil, ½ teaspoon of salt and the egg yolks, and stir in the cooked shrimps to make an egg/shrimp batter.
5. Put 2 tablespoons of oil in a cold wok, heat the oil to a medium-high temperature (about 170°C) and turn off the heat. Pour the egg/shrimp batter into the center of the oil.
6. Using a spatula, fold the batter from one side to the other repeatedly, always following the same direction, until 90% of the batter has coagulated into a soft scrambled egg. Place eggs and shrimps over the rice vermicelli.

Tips

1. Add a little heat if the batter does not coagulate properly.
2. Eggs will continue to cook in residual heat even after being put onto a plate.

生煎菜肉鍋貼
Pot Stickers

Serves *2* / Preparation *30* mins /
Marinate *15* mins / Cooking time *5* mins

Ingredients
150 g dumpling skin, large
150 g minced pork
300 g Chinese cabbage
1 tsp light soy sauce
¼ tsp salt
½ tsp sugar
½ tbsp corn starch
1 tsp sesame oil
2 tbsp Zhenjiang vinegar
1 tbsp shredded ginger

Method
1. Marinate minced pork with soy sauce, salt, sugar and 2 tablespoons of water for 15 minutes.
2. Blanch cabbage until soft, rinse and drain.
3. Chop cabbage, squeeze to remove most of the water, and mix with pork.
4. Mix in ½ tbsp corn starch and stir in sesame oil.
5. Place about 1 tablespoon of filling in the center of a piece of dumpling skin, brush some water on the edge of the skin and fold the skin to make a half circle. Seal the top by squeezing tightly. Start on one end of the half circle and seal by creating folds and squeezing the folds together. Repeat on the other end to complete the half-moon shape of the pot sticker. Finish making all the pot stickers.
6. Heat 1 tablespoon of oil in a non-stick skillet over medium heat and line the skillet with pot stickers. Add 60 ml of water (about 4 tablespoons) and bring to a boil. Cover the skillet, reduce to low heat and pan fry until all the water has evaporated. Continue to pan fry until a brown crust forms at the bottom of the pot stickers.
7. Serve with shredded ginger and Zhenjiang vinegar.

Serves *2* / Preparation *10* mins /
Marinate *20* mins / Cooking time *30* mins

Ingredients

1 pc Cantonese sausage
½ chicken, or 2pc chicken thighs
160 g rice
2 tbsp chopped garlic
1 tbsp ginger juice
1 tsp light soy sauce
1 tbsp corn starch
1 tbsp shredded ginger
2 stalks spring onion stems

Sauce for rice

1 tbsp dark soy sauce
2 tbsp light soy sauce
1 tbsp chicken broth or boiled water
½ tsp sugar
½ tsp sesame oil
Mix all ingredients into a sauce and bring to a boil.

Method

1. Wash sausage and slant cut into 1 cm thick slices. Cut spring onion stems into 5 cm lengths.
2. Discard chicken head and tail, cut chicken into pieces and marinate with soy sauce and ginger juice for 20 minutes, and mix in corn starch. Stir-fry chicken in 1 tablespoon of oil until half done.
3. Wash rice and drain dry, stir-fry chopped garlic in 1 tablespoon of oil until pungent, add rice and stir-fry together for about ½ minute.
4. Put rice into a casserole, add an appropriate amount of fresh water, and cook uncovered over high heat.
5. When most of the water has evaporated, put in sausage and chicken, and top with spring onion and ginger. Cover casserole and cook over low heat for 20 to 25 minutes.
6. Serve together with sauce for rice.

椰汁咖喱大蝦
Sautéed Prawns in Curry Sauce

Serves *2* / Preparation *10* mins / Cooking time *10* mins

Ingredients
300 g large prawns
2 tbsp corn starch
6 pc okra
4 cloves garlic
4 pc shallot
1 pc red cayenne pepper
3 pc dried chili pepper
1 tbsp curry powder/sauce
50 ml coconut milk
1 tbsp sugar
1 tbsp fish sauce
1 stalk basil

Method
1. De-vein prawns by inserting a toothpick behind the second abdominal segment to pick out the vein, then cut off antenna and legs using kitchen shears. Using a sharp knife, cut open the body along the abdomen, and dust the inside and outside of the prawns with corn starch.
2. Heat 4 tablespoons of oil in a wok and shallow fry prawns in batches to a golden brown. Remove prawns to drain oil.
3. Wash okra and cut off the stem ends.
4. Chop garlic, slice shallots, deseed red cayenne pepper and cut into two halves.
5. With the oil remaining in the wok, stir-fry garlic, shallot and dried chili peppers over medium heat until pungent. Stir in curry powder and add 2 tablespoons of water.
6. Put in okra and red cayenne pepper, and stir-fry together with the other ingredients in the wok.
7. Add prawns and coconut milk, sugar and fish sauce, and sauté to reduce the sauce.
8. Transfer to a plate and top with basil.

Tips

1. The use of dried chili peppers is to increase the tanginess of the dish and may be omitted.
2. Red cayenne pepper enhances the color of the dish but does not increase its tanginess.

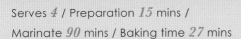

蜜汁叉燒
BBQ Pork

Serves *4* / Preparation *15* mins /
Marinate *90* mins / Baking time *27* mins

Ingredients
600 g pork shoulder

Marinade
6 tbsp sugar
1 tsp 5 spice powder
½ tsp Shajiang powder
2 tbsp Shaoxing wine
2 tbsp chopped shallot

1 tbsp salt
2 tbsp Hoisin sauce
½ tbsp light soy sauce
2 tbsp chopped garlic
1 tbsp ginger juice

Ingredients for honeyed sauce
3 tbsp maltose
1 tbsp mirin

3 tbsp sugar
1 tbsp boiled hot water

Mix and dissolve honeyed sauce ingredients in a small bowl and heat in a microwave oven for 30 seconds. It can also be put into a metal container placed over hot water.

Method
1. Clean pork and cut into strips about 2 to 2.5 cm thick.
2. Soak pork in 500 ml of water with 1 tablespoon of salt for 30 minutes, (the water enough to cover the pork) rinse, drain, and pat dry with kitchen towels.
3. Mix all the marinade ingredients and marinate pork for 1 hour. Turn over pork 2 to 3 times.
4. Skewer pork on metal skewers.
5. Preheat oven to 190℃. Cover a baking tray with a sheet of aluminium foil and place a BBQ wire mesh on top. Place pork on the wire mesh and roast for 16 to 17 minutes.
6. Brush pork with a coat of honeyed sauce, change the cooking method to broil and broil for 5 minutes. Turn pork over, brush with another coat of honeyed sauce and broil for another 5 minutes.
7. Remove pork to a plate and brush on one more coat of honeyed sauce.

Tips
1. Select the pork shoulder with some fat on it. When cut into strips, each strip of pork should also carry some fat.
2. Regardless of which part of pork is used, the sauce and timing remain the same.
3. We have replaced the rose flavoured wine used in traditional BBQ sauce with mirin to add a shiny sheen to the surface of the pork.
4. Take time when brushing on honeyed sauce to allow the heat of the pork to dissolve the thick sauce and coat the pork entirely.
5. A dip sauce may also be made to put on top of the pork. Simply filter the marinade sauce, add a little sugar and water, and bring to a boil.

蜜燒雞膶
BBQ Chicken Liver with Honeyed Sauce

Serves *4* / Preparation *10* mins /
Marinate *30* mins / Cooking time *20* mins

Ingredients
8 sets chicken liver (one set has 2 pieces of liver)

Ingredients for BBQ sauce
6 tbsp sugar
1 tsp 5 spice powder
2 tbsp Hoisin sauce
½ tsp Shajiang powder
½ tbsp light soy sauce
2 tbsp Shaoxing wine
2 tbsp chopped garlic
2 tbsp chopped shallot
1 tbsp ginger juice

Honeyed sauce ingredients
2 tbsp maltose
2 tbsp sugar
1 tbsp mirin

Utensil
4 pc bamboo skewers

Method
1. Remove fat from the surface of chicken liver and tear out any blood vessels. Rinse liver and drain.
2. Marinate liver with BBQ sauce for 30 minutes.
3. Place honeyed sauce ingredients in a bowl and add 1 tablespoon of hot water to dilute maltose.
4. Soak bamboo skewers in water for 10 minutes so that they will not burn during roasting.
5. Skewer chicken livers with bamboo skewers. Use two skewers for each string of four sets of liver.
6. Preheat oven to 190℃. Place a piece of aluminum foil on the bake pan and brush on a coat of oil. Put skewered livers on the pan.
7. Put the pan in the middle level (if available) in the oven and roast for 10 minutes.
8. Remove pan from the oven and brush livers with a coat of honeyed sauce. Rest for about 2 minutes to allow liver to cool off.
9. Set the oven to broil, and return bake pan with the livers to the top level of the oven and broil for 3 minutes.
10. Flip over the livers together with the bamboo skewers, brush livers with another coat of honeyed sauce, and broil for 2 minutes.

鹽烤馬友魚
Broiled Four Finger Threadfin

Serves *2 to 3* / Preparation *10* mins /
Marinate *1* hour / Cooking time *20* mins

Ingredients
1 pc four finger threadfin, about 500 g
½ tsp salt (for fish)
½ tsp salt (for fish tail)
¼ tsp ground white pepper
1 tbsp egg white

Method
1. Scale fish and clean.
2. Open up fish and cut open the vein that runs alongside the spine. Clean the vein with finger or a brush and rinse. Dry fish inside and out with kitchen towels.
3. Make 2 or 3 shallow cuts on each side of the fish, and marinate fish inside and out for 1 hour with ½ teaspoon of salt and ground white pepper.
4. Brush the tail fin with egg white and coat with salt to avoid burning the fin during baking.
5. Line a bake tray with aluminum foil, put on a wire mesh and brush mesh with a coat of oil. Put fish on the wire mesh.
6. Preheat oven to 200°C, put fish in the oven to bake for 8 minutes.
7. Flip over fish and bake for another 8 minutes.
8. Set oven to broil and broil fish for 4 minutes. Remove fish from the oven and serve.

蒜香金沙骨
Garlic Flavored Spareribs

Serves *2* / Preparation *20* mins / Cooking time *20* mins

Ingredients
300 g spareribs
1 tsp light soy sauce
3 tbsp chopped garlic
4 tbsp chopped shallot
¼ tsp salt
1 tbsp sugar
1 tsp corn starch
½ tbsp curry powder
½ tbsp ketchup
2 tbsp butter

Method
1. Chop spareribs into 6 to 7 pieces, each about 5 cm in length. Marinate with soy sauce, garlic, shallot, salt, sugar and 2 tablespoons of water for 15 minutes.
2. Mix in corn starch, curry powder and ketchup.
3. Put spareribs in a bake tray. Drizzle remaining marinating sauce on top of spareribs and brush on melted butter.
4. Pre-heat oven to 200°C, put in spareribs and roast for 10 minutes.
5. Turn over spareribs and roast for another 10 minutes until slight brown on the surface.

Baked Cauliflower in a Cheese Sauce

Serves *2* / Preparation *20* mins / Cooking time *30* mins

Ingredients
½ head cauliflower, about 250 g
2 slices ham
3 pc egg

Cheese sauce
2 slices cheese
2 tbsp butter
2 tbsp flour
250 ml milk
½ tsp salt

Topping
1 tbsp grated Parmesan cheese
1 tbsp bread crumbs

Method
1. Remove the stems of the cauliflower and retain the flowers. Blanch, drain, and put into a bake pan.
2. Chop ham. Boil the eggs, shell, and chop.
3. Chop cheese slices.
4. Melt butter in a pan over low heat, add flour gradually and stir to mix thoroughly to remove all lumps. Add milk and salt, and bring to a boil. Stir in chopped cheese and cook over low heat until all the ingredients are blended into a white cheese sauce.
5. Mix 1/3 of the cheese sauce with the cauliflower in the bake pan.
6. Distribute chopped ham and eggs evenly on top of cauliflower, and top with the remaining cheese sauce.
7. Sprinkle Parmesan cheese and bread crumbs on top of cheese sauce.
8. Preheat oven to 190°C.
9. Place cauliflower into the preheated oven, bake for 15 minutes, to be followed by broiling for about 4 minutes or until light brown.

Tips
1. Breakfast ham is used in this recipe. Other kinds of ham may also be used. If the ham is too salty, reduce the salt in the cheese sauce.
2. If the cheese sauce is too thick, add an additional 50 ml of milk or water.

香草焗雞
Roast Chicken with Herbs

Serves *4* / Preparation *15* mins /
Marinate *1* hour / Cooking time *45* mins

Ingredients
1 dressed chicken, about 1.2 kg
2 tsp salt
1.5 tbsp chopped garlic
1.5 tbsp mixed herbs (Italian or Provence)
2 tbsp olive oil
300 g potato
150 g carrot
200 g onion

Method
1. Clean chicken and remove head and neck. Chop off chicken feet 1 cm below the joint of the legs.
2. Brush chicken inside and out with salt and marinate for 1 hour.
3. Cut potatoes in half, slice carrots, and cut onions into large pieces
4. Mix garlic, herbs and 1 tablespoon of olive oil into an herbal sauce. Rub the inside of the chicken with half the sauce.
5. Insert fingers beneath the skin on the chicken breast and separate skin and meat, and spread ¼ of the herbal sauce on the meat. Massage with fingers gently to allow the flavor of the herbs to penetrate the meat.
6. Brush the remaining ¼ of the herbal sauce on the skin.
7. Sew the opening at the tail with needle and thread, and tie the legs with a cotton string to bring them together.
8. Fold the wings backwards next to the back of the chicken.
9. Preheat oven to 190°C.
10. Put chicken in a bake pan with potatoes, carrots and onions on the side. Drizzle 1 tablespoon of olive oil on the vegetables. Bake for 45 minutes.
11. Take chicken out of the oven, cut the cotton string that ties the legs, and remove the thread at the tail.
12. Pour out the juice from inside the chicken and use it as a dip sauce.

Serves *2* / Preparation *10* mins /
Marinate *1 to 2* days / Cooking time *15* mins

Ingredients
250 g cod
2 tbsp mirin
2 tbsp sake
1.5 tbsp Shiro miso
1.5 tbsp sugar

Method
1. Rinse fish, drain and pat dry with kitchen towels.
2. Heat mirin and sake in a small pot over low heat until all the alcohol has vaporized.
3. Add miso and mix with a spoon until all the ingredients are fully blended into a paste.
4. Put in sugar and stir until sugar is totally dissolved and mixed with the paste. Set aside to cool.
5. Put fish into a food storage bag, add paste, and seal bag after squeezing out the air.
6. Gently massage sauce into the fish with fingers. Refrigerate for 1 to 2 days.
7. Remove fish from the storage bag and wipe off excess sauce from the fish with fingers. Set aside to warm to room temperature.
8. Preheat oven to 230°C, place fish in a bake tray and bake for about 6 minutes. Turn fish over and bake for another 6 minutes. The fish can also be broiled, 4 minutes on each side until slight brown.

Tips

Flounder or sea bass can also be used in place of cod.

Ingredients

6 to 7 pc dried fish maw (about 50 g each piece)
50 g ginger slices
fresh water as needed

Method

1. Soak dried fish maw in fresh water for at least 8 hours or overnight until soft. Rinse.
2. Put fish maw and ginger into the electronic pressure cooker and add enough water to cover 4 cm above fish maw. Close and lock cover.
3. Set cooker to "low pressure" for 12 minutes and press start. Cooking is done when the cooker finishes de-pressurizing. For thicker fish maw, add a few more minutes.
4. Remove fish maw from the cooker and rinse with fresh water. The fish maw can now be used in other dishes, or put into individual food storage bags and stored in the freezer for later use.

冬菇燴花膠
Braised Fish Maw with Mushrooms

Serves *2 to 4* / Preparation *10* mins / Cooking time *15* mins

Ingredients
50 g dried fish maw
8 pc dried black mushroom
1 tbsp oyster sauce
½ tbsp corn starch
1 tbsp ginger juice
2 tbsp chicken broth
½ tbsp Shaoxing wine
½ tsp sugar
½ tsp salt
½ tsp sesame oil

Method
1. Prepare fish maw (see Page 140) and cut into 4 pieces.
2. Soften mushrooms in water, remove stems, and squeeze to remove the water. Mix mushrooms with oyster sauce and corn starch, and stir in 1 teaspoon of oil.
3. Put mushrooms in an electronic pressure cooker, add chicken broth and ginger juice, set to low pressure for 10 minutes, and press start to begin.
4. When the cooker has de-pressurized, open the cooker, add fish maw and stir in wine, sugar, and salt.
5. Close the cooker, set to low pressure for 5 minutes, and press start. When cooking is completed, stir in sesame oil and transfer to plate.

溏心鮑魚
How to Make Soft Core Abalone

Ingredients

10 pc Yoshihama abalone (30 heads per 600 g)
6 pc dried scallop
400 g pork spareribs
12 pc chicken feet
2 pc pork skin, about 7.5 x 15 cm each
2 slices ginger
50 g pressed cane sugar

Method

1. Wash abalones and soak abalones in fresh water overnight. Pour out water and clean abalones with a soft brush. Be sure to clean the sides and the mouth of the abalones.

2. Place abalones in a food storage box, and add fresh water to cover abalones by at least 1 cm. Cover box and refrigerate for 2 days. Save the water for later use.

3. Soak dried scallops with fresh water overnight. Save the water for later use.

4. Chop spareribs into several pieces, wash and clean chicken feet; scrape pork skin clean of all hair and fat, and then rinse. Blanch spareribs, chicken feet and pork skins for about 3 minutes and rinse with fresh water.

5. Put spareribs on the bottom of the electronic pressure cooker, and add, in order, abalones, chicken feet, scallops, ginger, pressed sugar and pork skins. Add water from soaking abalones and scallops and enough of fresh water to cover all the ingredients. Close and lock the cooker.

6. Set cooker to high pressure for 50 minutes, and press start to begin. When cooking is completed and the cooker de-pressurized, repeat the above procedure once more.

7. When the cooker changes to "keep warm", leave the abalones inside for at least another 30 minutes before unlocking and opening the cover to remove the abalones. The abalones are ready when they can be penetrated easily with a needle.

8. Put abalones on a rack and air for at least 2 hours for the softening

of the core. Abalones can then be placed in food storage bags and frozen if not immediately consumed.

9. Run the soup from the cooker through a sieve into a pot and reduce over low heat into an abalone sauce. Take care, however, to have enough sauce to cover the abalones completely when it is time to reheat the abalones. Do not add any salt to the sauce. Once the sauce has cooled off, put it into a food storage box and freeze to keep for later use.

When it time to serve the soft core abalone

1. Defrost abalone under room temperature if it is taken directly from the freezer. Put abalone into a bowl. Heat enough abalone sauce to cover the abalone completely and steam for 15 minutes.
2. Transfer the abalone to a plate without the sauce.
3. Heat the abalone sauce in a clean pot and thicken with 1 teaspoon of corn starch mixed with 1 tablespoon of water. Flavor with oyster sauce or cured ham flavoured sauce if needed.
4. Pour sauce over the abalone before serving.

How to make cured ham flavored sauce (if needed)

Marinate 5 g of Jinhua ham (or Yunnan ham) slices with 1 teaspoon of sugar, 1 tablespoon of Shaoxing wine and 2 tablespoons of water for 10 minutes. Steam over high heat for 15 minutes to become cured ham flavored sauce.

青咖喱牛面頰
Beef Cheek in Green Curry

Serves *2 to 3* / Preparation *15* mins / Cooking time *35* mins

Ingredients
1 pc beef cheek, about 400 g
200 g carrot
200 g potato
4 pc Thai eggplant
4 pc shallot, sliced
3 cloves garlic, sliced
3 tbsp Thai green curry paste
½ tsp salt
2 tbsp sugar
100 ml coconut milk

Method
1. Remove tendons and membrane from the beef cheek, rinse and drain.
2. Peel and slice carrots and potatoes; remove the stems of the Thai eggplants and cut each into two halves.
3. Put 1 tablespoon of oil in the electronic pressure cooker, set to auxiliary cooking for 10 minutes and press start to begin. Add shallot and garlic and stir until pungent. Put in curry paste, salt and sugar, and mix into a curry sauce.
4. Put in beef cheek, carrots, potatoes and Thai eggplants, and mix with curry sauce.
5. Move the beef cheek to the bottom of the cooker and covered by the vegetables.
6. Add 2 tablespoons of water along the inside of the cooker.
7. Close and lock the cooker, set to high pressure for 20 minutes, and press start.
8. Open the cooker when it has been de-pressurized. Take out beef cheek to cut into pieces and return to the cooker.
9. Add coconut milk to the cooker and mix with all the ingredients.

小碗薄切梅菜扣肉
Pork Belly with Preserved Mustard Hearts

Serves *2* / Preparation *20* mins / Cooking time *40* mins

Ingredients
150 g pork belly
75 g preserved sweet mustard heart
¼ tbsp dark soy sauce
1 tbsp sugar
½ tbsp Shaoxing wine
½ tbsp ginger juice

Method
1. Scrape pork belly clean of hair, rinse, and blanch for 15 minutes. Flush with fresh water until cool.
2. Soak mustard heart in fresh water for 5 minutes and rinse with water under the faucet. Squeeze water from the mustard hearts.
3. Chop mustard hearts after cutting off and discarding the firm part near the stem.
4. Marinate mustard heart with sugar, wine and ginger juice, and stir in 2 tablespoons of oil.
5. Cut pork belly into 2 mm thick slices and mix with dark soy sauce.
6. Line the bottom and sides of a 500 ml bowl with pork belly, add chopped mustard hearts and smooth out by pressing gently.
7. Put a steaming rack at the bottom of the electronic pressure cooker, add 125 ml of water, and place the bowl with the pork and mustard heart on top of the rack.
8. Close and lock cover, and set to high pressure for 40 minutes. Press start to begin.
9. Open the cooker when de-pressurization is completed, and remove the bowl. Cover the bowl with a small plate and drain the sauce into another bowl. Turn over the bowl to transfer the meat and mustard hearts to another plate and pour the sauce back on top.

Tips

If mustard heart with leaves are used, make sure each leaf is cleaned thoroughly.

清湯獅子頭
Stewed Meatball in Clear Broth

Makes *3* small bowls / Preparation *15* mins / Cooking time *20* mins

Ingredients

150 g minced pork shoulder
100 g fatty pork
50 g shrimp
½ tbsp oatmeal
1 pc small Chinese cabbage
½ tbsp ginger juice
½ tsp salt
⅛ tsp sugar
2 tsp corn starch
a pinch ground white pepper
chicken broth as needed

Method

1. Cut fatty pork into thin strips and then into small bits size of half a pea.
2. Shell shrimps and chop into a paste.
3. Crush oatmeal into a powdery form by hand.
4. Tear off and rinse cabbage leaves.
5. Put all the ingredients other than cabbage and chicken broth in a large bowl and mix by hand until fully blended. Make three meatballs.
6. Line three white ceramic or stainless steel containers with cabbage, add a meatball to each container and enough chicken broth to cover meatball.
7. Put a steaming rack on the bottom of the electronic pressure cooker, add 125 ml of water, and place the three containers into the cooker. Do not cover the containers.
8. Close and lock the cooker, set to high pressure for 20 minutes and press start to begin. The meatballs are ready to be served when the cooker has been de-pressurized.

Tips

1. If not using an electronic pressure cooker, simply seal the containers with aluminum foil and steam for 2 hours.
2. Fatty pork is essential for the meatball to have the right texture. Freezing the fatty pork first will facilitate cutting. Do not chop fatty pork.
3. Oatmeal helps to hold the meatball together. However, do not pick up the meat and slap it against a bowl as in the case of making ordinary meatballs as this will change the texture of the meatball.

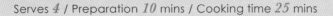

Serves *4* / Preparation *10* mins / Cooking time *25* mins

Ingredients
600 g beef plate
300 g beef tendon
600 g turnip

Condiments
3 tbsp Zhuhou sauce
2 pc star anise
1 tbsp Sichuan pepper
1 tbsp corn starch
30 g ginger slices
2 tbsp Shaoxing wine
2 tbsp oyster sauce
1 tbsp sugar
1 section lohan fruit

Method
1. Heat a large pot of water and blanch beef and tendons for 5 minutes. Rinse.
2. Cut beef into 4 cm squares, and tendons into 6 cm lengths.
3. Peel and cut turnip into large chunks. Roll the turnip slightly with each cut.
4. Put Sichuan pepper, star anise and lohan fruit into a spice pouch. Close or tie the pouch.
5. Add corn starch to the beef and mix by hand to coat each piece.
6. Mix beef, tendon and turnip with the remainder of the condiments and put into the electronic pressure cooker. Place the spice pouch at the bottom.
7. Set the cooker to high pressure for 20 minutes and press start to begin.
8. The beef, tendon and turnip are ready to be served once the cooker has de-pressurized.

Tips

Cooking with an electronic pressure cooker usually does not require putting in more water because the ingredients will release their juice naturally under pressure.

豬腳薑醋二人前
Vinegar with Pork Trotters and Ginger, for Two

Serves *2* / Preparation *10* mins /
Marinate ginger *30* mins / Cooking time *40* mins

Ingredients
2 pc egg
1 pc pork trotter
80 g ginger
1 tsp salt (for ginger)
300 ml sweet vinegar
(must be enough to cover pork trotter and ginger)

Method
1. Put eggs in a pot of cold water, heat the water to a gentle boil, change to low heat and remove eggs after three and half minutes. Flush eggs with cold water until cool. Peel eggs and soak eggs in sweet vinegar until they take on a nice brown color.
2. Clean pork trotter and cut into 6 pieces, blanch for 10 minutes and flush with cold water until cooled.
3. Peel and cut ginger into thick and long pieces. Crush ginger.
4. Marinate ginger with salt for 30 minutes. Rinse to remove the salt.
5. Place steaming rack in the electronic pressure cooker, and add 125 ml of water.
6. Put ginger and vinegar in a large bowl, and place the bowl on the steaming rack in the cooker. Close and lock the cooker, and set cooker to high pressure for 15 minutes. Press start to begin cooking.
7. When the cooker has de-pressurized, open the cover and put pork trotters into the bowl. Make sure there is enough vinegar to cover the trotters.
8. Close and lock cooker and set to high pressure for 20 minutes. Press start to begin.
9. Serve the eggs together with the pork trotters, ginger and vinegar.

Tips

1. It is better to use a container that is taller rather than wider to facilitate easier removal from the electronic pressure cooker. Less vinegar is also needed to cover the ingredients.
2. Eggs will be overcooked if cooked together with the vinegar and should be cooked separately.

番茄洋葱燴牛脷
Beef Tongue Bolognese

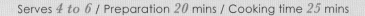

Serves *4 to 6* / Preparation *20* mins / Cooking time *25* mins

Ingredients

1 pc beef tongue (about 1.4 kg)
300 g tomato
300 g onion
300 g carrot
1 can tomato paste (170 g)
1 tbsp Italian or Provence mixed herbs
1.5 tbsp salt
2 tbsp sugar

Method

1. Pare off the skin of the tongue with a sharp knife (or request the meat seller to do so), wash, and cut off the tongue from its root.
2. Cut the tongue into 1 cm thick slices. Cut the root down the middle into two sides and cut each side into 1.5 cm thick slices. Soak both tongue and root in fresh water for 10 minutes, and drain.
3. Wash tomatoes and make a cross cut on the skin at the bottom of each tomato. Blanch tomatoes for 2 minutes and peel. Cut each tomato into 8 sections.
4. Peel and cut each onion into 8 sections.
5. Peel and cut carrot into 1.5 cm thick slices.
6. Put tongue and root at the bottom of the electronic pressure cooker, and top with onion and carrots.
7. Mix tomato paste with tomatoes, salt, sugar and herbs, and put on top of onion, carrots and tongue.
8. Close and lock cooker, set to high pressure for 20 minutes, and press start to begin cooking. The cooking is done when the cooker has de-pressurized.

Tips

1. The texture of the tongue and the root are different, therefore cuts of different thickness are used.
2. Tomatoes and onions release sufficient juice so that additional water is not needed when cooking with electronic pressure cooker.

Serves *2* / Preparation *5* mins / Cooking time *20* mins

Ingredients
10 g dried kelp
5 g dried seaweed
300 g spareribs
5 g ginger slice
½ tsp salt

Method
1. Soak dried kelp in fresh water for 30 minutes, clean, and cut into sections.
2. Rinse and drain spareribs.
3. Place spareribs and kelp together with ginger and 750 ml of water into the electronic pressure cooker.
4. Close and lock cooker, set to high pressure for 20 minutes, and press start to begin.
5. Open cover when cooker has been de-pressurized, add seaweed, and flavor with salt.

Tips

1. Kelp is available in supermarkets and wet markets.
2. Seaweed for sushi work best by simply stir into the hot soup.
3. If ordinary stock pot is used, bring the soup to a boil and reduce to low heat to simmer for 2 hours.

鮮人參雞湯
Chicken Soup with Fresh Ginseng

Serves *2 to 3* / Preparation *30* mins /
Soaking time *2* hrs / Cooking time *40* mins

Ingredients

1 dressed chicken or silkie chicken, about 1 kg
1 pc fresh ginseng (about 30 to 35 g)
120 ml glutinous rice
12 pc shelled gingko nut
8 pc shelled chestnut
3 pc jujube date
4 cloves garlic
½ tsp salt (for soaking rice)
1 tsp salt (for soup)
1 litre water

Method

1. Mix glutinous rice with ½ teaspoon of salt, add water equal to the quantity of rice, and soak for 2 hours. Rinse with fresh water and drain.
2. Clean the inside of the chicken, tear out lungs and kidneys, and remove head and neck. Cut off chicken feet 1 cm below the joint and save for later. Rinse chicken and drain.
3. Clean ginseng with a soft brush, rinse and cut into 2 pieces.
4. Pit and skin the jujube dates. Peel garlic.
5. Stuff chicken with glutinous rice, garlic and ginseng, and sew the opening at the tail with needle and thread.
6. Truss chicken by tying the two legs with a cotton string to bring them together and fold the wings backwards next to the back of the chicken.
7. Place chicken inside the electronic pressure cooker, add chicken feet, gingko nuts, chestnuts, jujube dates and salt together with water.
8. Close and lock cooker and set to high pressure for 40 minutes. Press start to begin.
9. When the cooker has de-pressurized, transfer chicken and soup to a large bowl and remove the thread and cotton string.

Tips

1. If a regular stock pot is used, boil 3 litres of water, put in trussed chicken (stuffed with glutinous rice, ginseng and others), cover and re-boil. Reduce to low heat and cook for 3 hours.
2. Fresh ginseng is available in Korean food stores. Do not wash ginseng until use. Fresh ginseng, wrapped and refrigerate, can be kept fresh for several days.
3. The soup will not be colored by the jujube dates once the skins are removed.

Serves *4* / Preparation *15* mins / Cooking time *25* mins

Ingredients

300 g pork (or beef)
4 cloves garlic
½ pc beet
150 g potato
150 g carrot
½ head cabbage
150 g onion
1 tbsp butter
1 can tomato paste (170 g)
1 pc bay leaf
1 tsp salt
sour cream

Method

1. Rinse and cut meat into 2 cm cubes, and put into the electronic pressure cooker together with peeled garlic. Add 1 litre of water and set cooker to high pressure for 10 minutes. Press start to cook into a meat broth.
2. While the meat broth is being made, cut beet, potato and carrot into 1.5 cm cubes.
3. Cut cabbage and onion separately into thick strips.
4. Stir-fry onion in butter until soft, and put into the meat broth in the cooker together with cabbage, beets, potato, carrot, tomato paste, bay leaf and salt. Mix well.
5. Set the cooker to high pressure for 10 minutes and press start to begin. The borscht is ready to be consumed after the cooker has de-pressurized.
6. Serve with sour cream.

花膠響螺頭燉雞湯
Chicken Soup with Fish Maw and Sea Conch

Serves *2* / Preparation *20* mins / Cooking time *25* mins

Ingredients
60 g dried fish maw
2 pc frozen sea conch meat, about 250 g total
½ chicken, about 500 g
150 g pork shin
10 g ginger slices
½ tsp salt

Method
1. Prepare fish maw (see page 140).
2. Defrost sea conch, remove visceral and the hard cover of the foot, clean and cut into 1 cm thick slices.
3. Remove the skin, head, neck and tail of the chicken, and tear out lungs, kidneys and membrane. Clean thoroughly.
4. Wash and cut pork into small chunks.
5. Put conch meat, chicken, pork and ginger into the electronic pressure cooker, and add salt and 750 ml of water.
6. Close and lock cooker. Set cooker to high pressure for 20 minutes and press start to begin.
7. When the cooker has de-pressurized, add fish maw to the soup and cook for another 5 minutes under low pressure.

Tips

If an ordinary stock pot is used, just follow steps 1 to 4, put all the ingredients together with salt and 750 ml of water into the pot, bring to a boil over high heat, and reduce to low heat to simmer for 3 hours. Add fish maw and cook for another 30 minutes.

香茅薑茶
Ginger Tea with Lemongrass

Makes *1* litre

Ingredients
3 stalks fresh lemongrass   40 g ginger slices
50 g rock sugar

Method
Rinse lemongrass, cut into sections and crush with the back of a chopper. Put lemongrass, ginger, rock sugar together with 1 litre of water into the electronic pressure cooker and set cooker to high pressure for 5 minutes. Press start to begin. When the cooker has been de-pressurized, pour out and filter the ginger tea. The tea can be served hot or chilled.

Tips
If an ordinary stock pot is used, put lemongrass and ginger together with 1.5 litres of water into the pot. Bring to a boil, reduce to low heat to cook for 1 hour. Add rock sugar and cook until the sugar is dissolved. Filter tea and serve hot or chilled.

黑木耳紅棗茶
Black Fungus Tea with Jujube

Makes *1* litre

Ingredients
10 g white back fungus   8 pc pitted jujube date halves
3 slices ginger   40 g rock sugar

Method
Soak white back fungus in water for 1 hour, rinse and tear into smaller pieces. Put fungus, jujube dates, ginger, rock sugar together with 1.25 litres of water into the electronic pressure cooker, set cooker to high pressure for 15 minutes and press start to begin. When the cooker has been de-pressurized, pour out and filter the tea before serving.

Tips
If ordinary stock pot is used, put fungus, jujube dates, rock sugar, ginger and 1.5 litres of water into the pot, and bring to a boil over high heat. Reduce to low heat and simmer for 2 hours. Pour out and filter the tea before serving.

Makes *1* litre

Ingredients
40 g sweet and bitter apricot kernels (sweet apricot kernels to bitter apricot kernels ratio 4:1)
2 pc pear
40 g rock sugar

Method
1. Rinse and drain sweet and bitter apricot kernels.
2. Wash pears, and cut into four pieces without peeling. Discard core.
3. Place pears, sweet and bitter apricot kernels and rock sugar together with 1 litre of water into the electronic pressure cooker, set to high pressure for 15 minutes and press start to begin.
4. The drink is ready to be consumed after the cooker has de-pressurized.

Tips

This can also be made without using an electronic pressure cooker. Simply put all the ingredients together with 1 litre of water into a deep container, seal tightly, and steam over high heat for 1 hour.

糯米蒸排骨
Steamed Glutinous Rice with Spareribs

Serves *2-3* / Preparation *10* mins /
Marinate *15* mins / Cooking time *1* hr

Ingredients
250 ml glutinous rice
200 g spareribs, cut in pieces
½ tsp salt
2 cloves garlic
1 stalk spring onion
1 pc fermented bean curd (small)
1 tbsp light soy sauce
1 tsp cooking wine
½ tsp sugar
½ tsp corn starch

Method
1. Rinse and drain rice.
2. Chop garlic and dice spring onion. Mash bean curd and blend with 2 tablespoons of water into a bean curd sauce.
3. Rinse spareribs and marinate with bean curd sauce, soy sauce, wine, sugar and caltrop starch for 15 minutes.
4. Put rice together with salt and water into the electronic rice cooker. The amount of water should be about 40 ml less that the amount normally used. Select white rice and delicious settings and press start to begin.
5. When water begins to boil, put spareribs evenly on top of rice.
6. When cooking is done, do not open the lid for another 5 to 10 minutes. Remove spareribs, add spring onion and stir to mix rice before serving.

Tips
1. Glutinous rice needs less water than ordinary white rice. Spareribs will also yield additional juice. This is why we recommend using about 40 ml less water.
2. Leaving the spareribs and rice in the cooker for a few more minutes after cooking is finished, it is to ensure spareribs will be cooked thoroughly.

Serves *2-3* / Preparation *15* mins / Marinate for fish *15* mins /
Cooking time *normal rice cooking time*

Ingredients

160 g salmon
2 tsp mirin
2 tsp light soy sauce
4 pc asparagus
240 ml rice
½ tsp salt
1 tbsp olive oil
8 pc morel mushroom
chicken broth/water

Method

1. Place salmon in a food storage bag, add mirin and soy sauce, and seal. Massage fish with fingers through the bag to ensure that fish is completely covered by sauce. Marinate for 15 minutes.
2. Cut off and discard the tougher part of the asparagus near the thicker end.
3. Rinse and drain rice.
4. Soak morel mushrooms in 60 ml of water for 10 minutes, snip open mushrooms with a pair of scissors and clean the inside. Remove mushrooms, and filter mushrooms water for later use.
5. Put rice, salt and olive oil into a rice cooker, add mushroom water and the appropriate amount of chicken broth (or water), and press start to begin to cook rice.
6. Add mushrooms to the rice when water begins to boil, and continue to cook until most of the water has evaporated (about 5 to 10 minutes before rice is fully cooked).
7. Add salmon to the rice with the asparagus on the side. Continue to cook until rice is fully cooked.
8. Keep the lid of the rice cooker closed for another 5 minutes before serving to allow salmon to continue to cook in residual heat.

Tips

The amount of chicken broth or water together with the mushroom water is the same as the normal amount of water used for cooking rice, and should be adjusted based on the kind of rice and rice cooker used.

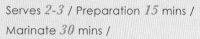

簡易海南雞飯
Hainan Chicken Rice

Serves *2-3* / Preparation *15* mins /

Marinate *30* mins /

Cooking time normal rice cooking time plus *10* mintues

Ingredients
½ chicken, about 700 g
320 ml white rice
2 tsp salt
1 tbsp ginger juice
½ tbsp Shaoxing wine
½ bulb chopped garlic
1 stalk lemongrass
100 ml coconut milk

Method
1. Clean chicken and marinate with 1.5 teaspoon of salt, ginger juice and wine for 30 minutes. Turing over chicken once after about 15 minutes.
2. Rinse and drain rice.
3. Heat 2 tablespoons of oil and stir-fry garlic until pungent. Add rice and ½ teaspoon of salt, and stir to mix thoroughly. Transfer rice to the electronic rice cooker.
4. Remove the outer skin of the lemongrass, rinse, and squash with a chopper. Cut lemongrass into 4 sections and put on top of rice.
5. Add water to the cooker. The amount of water added should be about 40 to 50 ml less than the ordinary amount. Select "Delicious" or slow cook and start the cooker.
6. When the water in the pot has come to a boil, open the lid, put in chicken with the skin facing upward, and drizzle the marinade sauce on top of chicken. Close the cooker and continue to cook.
7. Do not open the cooker when the cooking cycle has completed, but leave the chicken and rice to cook in residual heat for another 5 to 8 minutes.
8. Open the cooker and remove chicken and lemongrass. Stir in coconut milk, close the cooker and bake for another 5 minutes to allow the coconut flavor to penetrate the rice.
9. Cut chicken to pieces and serve together rice.

Tips

1. Cooking Hainan chicken rice using this method requires a sufficient quantity of rice to allow for enough time to fully cook the chicken. If a smaller quantity of rice is used, increase the time to 10 minutes before opening the cooker when the cooking cycle has completed.
2. 320 ml of rice equals 2 measuring cups of the rice cooker.
3. Reducing the amount of water for cooking rice is because the juice from the chicken will be released to the rice. The actual amount of water will depend on the type of rice use.
4. This recipe is for the kind of electronic rice cooker that can be opened during the cooking cycle. It is not appropriate for the traditional rice cooker.

Shanghai Vegetable Rice

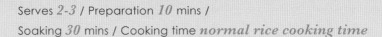

Serves *2-3* / Preparation *10* mins /
Soaking *30* mins / Cooking time *normal rice cooking time*

Ingredients
240 ml white rice
100 g Shanghai salted pork
30 g fatty pork
100 g Shanghai brassica
2 tbsp chopped garlic
1 tsp salt

Method
1. Rinse and drain rice.
2. Soak salted pork for 30 minutes to reduce its saltiness. Take out and cut into ½ cm thick pieces.
3. Cut fatty pork into small pieces.
4. Rinse and blanch vegetables, soak in cold drinking water until cool. Drain and chop vegetables.
5. Heat 1 teaspoon of oil in the wok and shallow fry fatty pork over low heat until crispy to extract lard. Remove pork crisp and save for later use.
6. Stir-fry chopped garlic in lard until pungent, add rice and stir for about 30 seconds.
7. Transfer rice to the rice cooker, add salt and the normal amount of water. Press start to begin to cook rice.
8. Add salted pork when most of the water has evaporated.
9. Stir in chopped vegetables when rice is fully cooked. Top with pork crisp when served.

Tips

1. Shanghai salted pork is available in stores selling Shanghai foodstuff. We suggest getting a piece with more fat on it as the fat will give out a richer taste.
2. Do not overcook the vegetables when blanching as they will be cooked further when mixed with hot rice.

皮蛋瘦肉粥
Congee with Salted Pork and Preserved Eggs

Serves *3 to 4* rice bowls / Preparation *10* mins /
Marinate *4* hrs / Cooking time *2* hrs

Ingredients
300 g pork
1 tsp corn starch
2 pc preserved eggs
80 ml white rice
1 tbsp salt

Method
1. Rinse and cut pork into slices, and marinate with salt for about 4 hours. Rinse away salt and mix with corn starch just prior to cooking.
2. Peel and cut each preserved egg into 8 pieces.
3. Rinse and drain rice. Put rice into the electronic rice cooker together with 800 ml of water, cover cooker, select white rice and congee, set time to 2 hours and press start to begin.
4. 10 minutes before congee is done, open the lid of the cooker, stir in pork and eggs, close cooker and continue to cook until congee is done.
5. Allow the congee to rest in the cooker for another 10 to 15 minutes before opening the cooker.

Tips

1. Fresh rice should be used as aged rice lose its starchiness over time. Thai and Australian long grain rice at a 1:1 ration works well. Rice from China's northeast, Taiwan or Japan works even better.
2. Do not add salt to the congee without tasting first as the pork had already been salted.
3. Keeping the congee in the closed cooker 10 more minutes allows time for the congee to become smoother. Making congee in an ordinary pot should also allow 15 minutes of "smooth out" time.

新一代電器廚具

一個高效、方便、環保的電器化廚房，
讓您增添煮食的新樂趣，並為您的至愛帶
來無限美食驚喜！

IH電磁爐

磁應煮食爐，香港人簡稱為電磁爐，有產品稱為IC電磁爐Induction Cooker，日本則稱為IH電磁爐Induction Hub。

我們在家中使用電磁爐無火煮食已經很多年，我們在香港出版的十多本食譜書，書中幾百個菜式的圖片，都是用電磁爐烹調出來的。現在我們已完全習慣用無火煮食，火候比傳統爐具更容易控制，加熱更快速，而且更節能。如果你覺得每月燃料費太昂貴，改用電磁爐煮食，每月可節省燃料費至少30%（數據由天祥公證行有限公司提供），既經濟又環保。

很多人都以為用電磁爐煮中菜會火候不夠，其實火力夠不夠，與你見到的火光大小無關，這些火光一部分是在空氣中燃燒所以被見到，而因此導至廚房溫度升高。而電磁導熱只會令接觸的鍋具產生熱量，是通過鍋具百分百地發揮功能，並沒有燃燒空氣，傳熱效率高達90%（數據由天祥公證行有限公司提供），有效地減低煮食產生的高溫及焦油，廚房溫度比採用明火煮食的溫度平均低4-5℃，最適合我們這些經常下廚的人。

用電磁爐煮食更適合有小朋友和老人家、寵物的家庭，備有各種安全功能：預設時間掣、自動斷電、剩餘熱力顯示燈、防止過熱保護等裝置，不怕忘記熄爐或空燒鍋具，使用得安心又放心。

IH金鑽西施電飯煲

米飯是亞洲人最重要的主食，生活中總離不開吃米飯，而電飯煲就是每一家人必備的廚房電器。發明電飯煲的是日本人，因為他們對烹煮米飯最為講究。1959年第一款日本電飯煲引入香港市場，1967年西施電飯煲面世。隨着時代的轉變，現在進入了磁應電飯煲的年代，粒粒飯芯，軟糯香腍，令米飯變成美食！

自1988年起，Panasonic首次將磁應加熱技術結合電飯煲，憑藉革命性的IH磁應技術，屹立電飯煲市場領導地位。IH金鑽西施電飯煲具自動浸米功能，並擁有5段IH強大火力，透過內鍋自身發熱，令米粒充分受熱受水，亦令米飯產生特有的米香。煮飯中途更可開啟飯鍋放入其他食材，輕輕鬆鬆地煮出香噴噴的海南雞飯和各式煲仔飯。

日本製造的 Panasonic HB 系列金鑽西施煲，幻影黑晶外觀不單型格高貴，而且機身小巧，比一般IH電飯煲體積少20-30%*，是現代廚房必不可少的電飯煲。此外，HB系列的內膽是『黃金陶鑽鍋』，高效傳熱，而中空的陶瓷斷熱層，有效封存IH熱量，保存米飯滋味。鍋內側採用「鑽石高導熱塗層」，產生大量熱對流細泡，米粒吸水更容易。作為IH技術的領導者，HB系列備有Panasonic專利的頂蓋IH技術，連同機身內藏4個IH電磁加熱線圈，全面覆蓋形成5段立體IH火力，全方位均勻受熱，讓熱氣經過米粒間一粒粒加熱，使米飯變得更飽滿。

 * 與Panasonic IH電飯煲作比較

多功能蒸氣焗爐

　　越來越多人愛追求健康飲食，更喜歡於家中親自炮製佳餚，全新的 Panasonic My Chef多功能蒸氣焗爐集多功能於一身，具備7種烹調組合，純蒸氣、蒸氣烤焗、蒸氣微波、單面燒烤、雙面燒烤、熱風對流烤焗、微波烹調，其中的蒸氣烹調有效保存食物的水分、鮮味及營養，讓你隨時在家中煮出健康、美味又有營養的美食。

　　全新的NN-CS894B多功能蒸氣焗爐，採取以純蒸氣烹調食物，無需使用保鮮紙，亦能保存食物中的水分，防止肉汁流失，保持食物鮮味；同時令食物更有效吸收熱能，將食物多餘的脂肪溶解排出。此爐特有『波浪紋燒烤盤』，燒烤盤上凹凸坑紋可將煮食時所產生之油分與食物隔開，減少食物的油膩感，更加健康，讓你在家中享受新鮮健康的美食。特強噴注式蒸氣設計，使食物迅速均勻加熱，令蒸煮效果更出色。此外，使用蒸氣解凍相比於室溫下解凍食物，更加快捷、衛生及健康。

　　NN-CS894B『變頻式』火力調較功能，能令加熱過程中的火力及溫度均維持於穩定的水平，令食物快捷均勻受熱。內置50款自動食譜烹調，並特設40℃烤焗功能，適用於發酵麵糰，無論是家庭主婦或上班忙碌一族同樣適合使用。電子輕觸式操控，操作簡易，輕鬆炮製出特色美食。另配備ECONAVI智慧節能功能，有效地節省能源；同時，大大提高煮食效率，更具自動清洗功能，清潔方便又簡單，帶來不同凡響的快捷煮食體驗。

電子高速煲

　　傳統的壓力煲，雖然烹調更快捷，但畢竟令人感覺到有點難以控制和有心理壓力。新一代的電子高速煲完全改變這種形像，以液晶體顯示屏控制，使用起來方便、安全、效能精確，功能一目了然，操作得心應手。從簡簡單單的燜牛腩、煲湯，以至發花膠、燜吉品乾鮑都全無難度，從此每天都可在家中烹調矜貴菜式，成為廚藝高手，我們誠意推介！

　　Panasonic電子高速煲最新推出型號SR-PG501（5公升）／ SR-PG601（6公升），備有嶄新設計安全鎖，設計時尚美觀，烹調美食更輕而易舉。Panasonic電子高速煲採用先進高壓烹調原理，熱能令鍋內的氣壓上升(最大壓力為98kPa)，加速溫度提升（最高溫度達120℃），能大大縮短烹調時間，比傳統烹調節省能源。而且密封設計的內鍋更能鎖住食物的原味和營養，避免在烹調中流失，使菜餚保持原汁原味！

　　Panasonic電子高速煲採用微電腦控制設計，可根據不同食材和煮食效果，選擇高、中或低壓烹調，和1-59分鐘的烹調時間。烹調結束後自動保溫，無需「睇火」，方便易用。更新增「手動排氣」按鈕，可更進一步確保鍋內壓力完全釋放，配備調壓閥、防堵塞設計、外蓋鎖及壓力顯示竿，設計至臻完美，安全可靠！

電焗爐

　　熱愛烹飪的人，家中都有一個烤焗爐，可以簡單快捷烹調美味的菜式。Panasonic最新推出NB-H3200電焗爐，烹調火力高達1500W，配合32公升超大容量，能一次烤焗大份量的食材，如烤雞或焗薄餅，製作任何食物得心應手。

　　NB-H3200擁有上下獨立控溫（70℃-230℃）、熱風對流、麵糰發酵、旋轉燒烤功能，滿足多種烘焗需要。無論是烹調達人或是廚房新手，都是入廚的必備之選。

　　NB-H3200電焗爐的上下火雙重控溫功能，可隨您的喜好與不同食材的需求設定溫度。調整上火溫度，讓烤焗食物輕鬆達到色澤金黃；烘焙蛋糕、麵包及加熱食材，可獨立設定下火溫度，令口感更美味更富層次。

　　如喜歡在家中自製麵包，Panasonic電焗爐NB-H3200更是您的最佳助手，它配備了發酵功能，讓麵糰維持在30℃-50℃的最佳發酵環境，配合使用Panasonic麵包機製作麵糰，便能滿足發酵與烘焙需要，讓您全面發揮創意，輕鬆創作各式各樣的麵包，如菠蘿包、雞尾包及牛角包等。

　　NB-H3200電焗爐內置的立體熱風對流能使熱力均勻地分佈，有效確保爐內的溫度均勻一致，烘焗各式各樣的食物同樣出色，讓您輕鬆烹調食物，享受真正烤焗的樂趣。採用360度旋轉燒烤，全面均勻受熱，琺瑯烤盤導熱更均勻且耐高溫，用後易於清洗。爐內更備有耐高溫的照明，食材清晰可見，讓您更易掌握烹調時間。

手提攪拌機

在香港這個寸金尺土的地方，喜歡烹飪的您希望盡收天下「兵器」一展廚藝，但家中位置有限，相信面對許多產品只能感到望門輕嘆。這個時候，一個輕巧而實用的手提攪拌機就最適合您了。在本書中介紹的班尼迪蛋，用上了手提攪拌機，做荷蘭醬就容易得多了。

Panasonic手提攪拌機MX-SS40配備攪拌棒、切碎器刀片、切絲及切片圓盤刀片和打蛋器等配件，一次過滿足您攪拌、切碎、切片或切絲和打蛋等多個需要。不銹鋼攪拌棒可直接放入容器中攪拌，攪拌西湯和果汁更快捷！配合1250毫升等大切碎器碗使用，以後無論是攪碎蔬菜、果仁、肉類，或是將蔬果及芝士等切片切絲也方便衛生！加上切碎器碗底部配有防滑墊，使用時更安全！

MX-SS40機身纖巧，易於掌控的手柄有助您在不同情況下靈活發揮烹飪創意，600瓦特高動力馬達可按需要調節轉速，絕對是您的廚房好幫手！

作者簡介

陳家廚坊

方曉嵐、陳紀臨夫婦，傳承陳家兩代的烹飪知識，對飲食文化作不懈的探討研究，是香港暢銷的食譜書作家。其十多本著作，內容豐富實用，文筆流麗，深受讀者歡迎。作品更在台灣多次出版，現正為國內和歐洲著名出版社編著中、英文食譜書，內容包括全國各地的地道菜式。

鳴謝

www.panasonic.hk

中華電力有限公司

馬廖千睿女士

伍文海先生

黎詩思女士

如有查詢，請登入：

f 陳家廚坊讀者會

或電郵至：
chanskitchen@yahoo.com

my COOKey
myCOOKey.com

美 食 · 簡 易 快　*Gourmet Cooking Made Easy*

作者　Author
陳家廚坊　Chan's Kitchen
方曉嵐 • 陳紀臨　Diora Fong • Keilum Chan

策劃/編輯　Project Editor
Catherine Tam

美術統籌及設計　Art Direction & Design
Amelia Loh

攝影　Photographer
Imagine Union

剪片　Film Production
Man Lo
Charlotte Chau
Wing Yeung
Ng Ming Wai

出版者　Publisher
萬里機構 • 飲食天地出版社　Wan Li Book Co LTD
香港鰂魚涌英皇道1065號東達中心1305室　Rm 1305, Eastern Centre, 1065 King's Road, Quarry Bay, Hong Kong
電話　Tel:　2564 7511
傳真　Fax: 2565 5539
網址　Web Site: http://www.wanlibk.com
http://www.facebook.com/wanlibk

發行者　Distributor
香港聯合書刊物流有限公司　SUP Publishing Logistics (HK) Ltd.
香港新界大埔汀麗路36號　3/F., C&C Building, 36 Ting Lai Road,
中華商務印刷大廈3字樓　Tai Po, N.T., Hong Kong
電話　Tel:　2150 2100
傳真　Fax:　2407 3062
電郵　Email: info@suplogistics.com.hk

承印者　Printer
美雅印刷製本有限公司　Elegance Printing & Book Binding Co., LTD

出版日期　Publishing Date
二〇一五年十一月第一次印刷　First print in November 2015

萬里機構

萬里 Facebook